The Best American Science and Nature Writing 2016

GUEST EDITORS OF
THE BEST AMERICAN SCIENCE
AND NATURE WRITING

2000 DAVID QUAMMEN
2001 EDWARD O. WILSON
2002 NATALIE ANGIER
2003 RICHARD DAWKINS
2004 STEVEN PINKER
2005 JONATHAN WEINER
2006 BRIAN GREENE
2007 RICHARD PRESTON
2008 JEROME GROOPMAN
2009 ELIZABETH KOLBERT
2010 FREEMAN DYSON
2011 MARY ROACH
2012 DAN ARIELY
2013 SIDDHARTHA MUKHERJEE
2014 DEBORAH BLUM
2015 REBECCA SKLOOT
2016 AMY STEWART

The Best American Science and Nature Writing™ 2016

Edited and with an Introduction by Amy Stewart

Tim Folger, Series Editor

A Mariner Original

HOUGHTON MIFFLIN HARCOURT

BOSTON • NEW YORK 2016

www.hmhco.com

ISSN 1530-1508
ISBN 978-0-544-74899-6

Printed in the United States of America
DOC 10 9 8 7 6 5 4 3 2 1

Contents

Foreword

IF, BY CHANCE, this anthology is read a century from now, what might those future readers make of it? First they would need to place the book in the context of its era, a pivotal one for all the world's inhabitants, human and nonhuman alike. Perhaps they'll recall that 2015 was the year our civilization established a grim new benchmark: the concentration of carbon dioxide in our planet's atmosphere reached 400 parts per million for the first time in recorded history, a level not seen since at least two million years ago, during the Pliocene epoch. Earth might as well have been another world then. Our own species, *Homo sapiens,* had not yet evolved; the average global temperature was five to seven degrees warmer; sea levels were anywhere from 16 to 130 feet higher than today.

In what sort of world will those future readers dwell? That depends on us and the choices we make over the next decade. Will we finally end our dependence on fossil fuels and avoid disaster? Will we be praised for our foresight or cursed for our shortsighted selfishness? Will historians in the 22nd century condemn our reckless dismissal of the increasingly urgent warnings of our best scientists? This much is certain: if we fail to act, it won't be because we didn't understand the magnitude of the threat.

As I began writing this foreword, James Hansen, the eminent Columbia University climate scientist, discussed the crisis in a video released to coincide with the publication of a startling new paper. The paper, which Hansen wrote with 18 coauthors from around the world, makes for sobering reading. He and his colleagues argue that we are perilously close to irreversibly dooming our descendants to the most catastrophic effects of climate change. "Have we passed a point of no return?" he asked in the video. "I doubt it, but it's conceivable." He added: "We are in a position of potentially causing irreparable harm to our children, grandchildren, and future generations. This is a tragic situation because it is unnecessary. We could already be phasing out fossil fuel emissions." All that's lacking is a commitment to action.

Hansen, who first testified before Congress—*in 1988*—that we were on a dangerous path, said it's very likely that the climate is changing even faster than computer models have been predicting. Specifically, the severity of future storms and the extent of sea-level rise later in this century may disastrously outpace current forecasts, according to his recent paper. If the seas rise by several meters—which can't be ruled out—the world's coastal cities would have to be abandoned. Imagine the loss of London, Shanghai, Tokyo, New York, Miami . . . And yet, given the enormous stakes, we haven't made much progress since Hansen's initial warnings nearly three decades ago. Greenhouse gas emissions are still rising, and so are the oceans.

Here's one disheartening gauge of our society's concern: How much time did the nation's leading nightly news programs—on ABC, NBC, CBS, and Fox—devote to the subject of climate change last year? It is, after all, the most pressing issue facing humanity. Without concerted international action, crops will fail; refugees will flee—*are* fleeing—flooded and drought-stricken lands; the extinction of species will accelerate. So, a big story, one worthy of serious, sustained reporting. There was even a religious angle when Pope Francis urged the world's leaders to act. Care to hazard a guess now about the coverage? Twenty minutes a week? Two hours a month? The correct answer: 146 minutes—for the entire year, for all the networks combined. Not even 3 minutes a week. Quarterback Tom Brady's "deflategate" imbroglio received twice as much airtime. Only one network, Fox, upped its "reporting" on the issue from the previous year—with a parade of talking heads denying the reality of climate change.

Fortunately, some of the country's best journalists *don't* ignore the most important story of our time. In "The Will to Change," Robert Kunzig covers Germany's effort to engineer "an epochal transformation" that will, if successful, slash the nation's planet-warming carbon emissions by 40 percent in 2020 and by 80 percent in 2050. If a cloud-shrouded, industrialized northern European country can end its long use of coal and other fossil fuels, maybe the rest of the world can, too.

Should the United States, China, and other nations fail to follow Germany's lead, there will be many more stories like Elizabeth Kolbert's "The Siege of Miami" and Gretel Ehrlich's "Rotten Ice." Large parts of Miami now regularly flood, Kolbert writes. The influx of salt water has started to threaten Florida's freshwater aquifers. If the eight-inch rise in sea level over the last hundred years has caused such problems, what will be left of Miami at the century's end, when seas will likely be at least three feet higher still? A significant part of that increase will come from Greenland's melting glaciers, which every year add some 50 billion tons of water to the oceans. Ehrlich, who has spent many months living in Greenland, gives a firsthand account of what it's like to travel by dogsled in the new, melting Arctic.

But Amy Stewart, this year's guest editor, hasn't limited her selections to climate change. She has given us a collection of stories that range far and wide, from the creation of a 15,000-page mathematical proof (Stephen Ornes's "The Whole Universe Catalog") to the life-threatening hazards of working in nail salons (Sarah Maslin Nir's "Perfect Nails, Poisoned Workers"). Ornes's piece is a rarity: an exciting story about mathematics. Only a few aging mathematicians understand the Brobdingnagian proof known as the Enormous Theorem, and there's a real risk that they will die before passing on their expertise to a new generation of savants. Nir's story features outstanding investigative reporting of a seemingly benign urban workplace. Manicurists—most of whom are poorly paid immigrants—are routinely exposed to a "toxic trio" of chemicals linked to miscarriages, cancer, and other painful maladies.

One of the other stories Amy has chosen—Kea Krause's "What's Left Behind"—describes how the solution to an intractable problem might lie in a most unexpected place: the heavily polluted waters of an abandoned copper mine. And then there's Amy Leach's luminous "The Modern Moose," which will make the day of anyone who reads it. I don't want to spoil the other surprises that you'll find in this anthology—our guest editor will have more to say about her selections in the next few pages—so I'll stop here, and invite all readers, present and future, to dive in.

With work on this year's anthology now over, I'm already gathering candidates for the 2017 edition. Do your part! I try to read widely, but without the help of many thoughtful readers, writers, and editors from around the world, I would miss some very good stories. Nominate your favorites for next year's anthology at http://timfolger.net/forums. I encourage writers to submit their own stories. The criteria for submissions and deadlines and the address to which entries should be sent can be found in the "news and announcements" forum on my website. Once again this year I'm offering an incentive to enlist readers to scour the nation in search of good science and nature writing: send me an article that I haven't found, and if the article makes it into the anthology, I'll mail you a free copy of next year's edition. Maybe our next guest editor will sign it, too. I also encourage readers to use the forums to leave feedback about the new collection and to discuss all things scientific. The best way for publications to guarantee that their articles are considered for inclusion in the anthology is to place me on their subscription list, using the address posted in the "news and announcements" forum.

I'd like to thank Amy Stewart for selecting such a diverse collection of stories for this year's anthology. As in years past, I'm very grateful to Naomi Gibbs and her colleagues at Houghton Mifflin Harcourt, who are responsible for the entire series of Best American anthologies. And finally, here's to many more years spent on a habitable and fair world with my beauteous wife, Anne Nolan. If I had to live in a Pliocene climate with anyone, it would be with her.

TIM FOLGER

Introduction

SOME OF YOU are going to skip this introduction. Hey, I'm not judging—but I do want to catch your eye before you go. There were two extraordinary essays published in 2015 that you won't find in this collection, but the only reason they didn't make the cut is because of length. Combined, they would have taken up almost half the book and displaced many other worthy pieces. But you must know about them.

You've probably heard that DNA evidence is being used to exonerate people who have been wrongly convicted of crimes, thus exposing the incompetence, impoverishment, and bias that plague the criminal justice system. DNA evidence is setting innocent people free, although not nearly enough of them. People are still behind bars who shouldn't be.

But that's biology. What about the science of arson? In her impeccable essay "Playing with Fire," which ran in *The Intercept* on February 24, 2015, Liliana Segura tells the harrowing story of a man convicted of first-degree murder on the basis of a flawed interpretation of the marks left behind when a house burned down. It turns out that certain kinds of burn patterns have long been considered unassailable proof that an accelerant had been poured or spilled around a room. However, we now know that the same marks can be made when "flashover" occurs—natural combustion caused by a buildup of radiant heat in a room.

Ms. Segura recounts the handful of arson convictions that have been overturned after a fresh review of the evidence, and speaks to the National Academy of Sciences' efforts to call attention to the outdated techniques still being employed by fire investigators and the faulty conclusions they draw from them. Her piece serves as a chilling reminder of the number of people who may have been convicted due to obsolete science, many of whom are still waiting for their cases to be reviewed.

"Playing with Fire" points to a broader theme that connects the best science writing of 2015. Recent advances in biology, ecology, physics, medicine, and engineering are bringing us closer than ever to solving the earth's and humanity's most pressing problems. But there are still enormous gaps between the problems and the solutions. In many cases, those gaps are the result of poverty, inequality, and prejudice.

Great science writing bridges that gap. It lets us—the readers, the citizens, the voters—understand what's possible. It shows us that even the most horrifying, human-made atrocities in our world (elephant poaching, poisoned water supplies, urban violence) can be understood and perhaps even solved through human ingenuity. It gives us something to advocate for. It unites us and inspires us.

I'll get back to that theme in a minute, but first I must introduce you to the other astonishingly good essay that couldn't fit between these pages. Paul Ford's "What Is Code?" appeared on Bloomberg.com on June 11 last year, and shortly thereafter the Internet exploded with joy.

I am not the only person to love this essay, in other words. This was a case of science writing going viral for all the best reasons. You might think that an extremely long article that attempts to explain to the general public how computer code works would be abysmally, painfully, soul-crushingly dull. But you would be wrong. This thing is amazing.

At just over 30,000 words, it exceeded our page limits. But that's just as well, because this is not a story that you want to read on paper anyway. It was built to be read onscreen, where it makes the best possible use of animation, mouse-over effects, and other clever interactive features that allow you, the reader, to test-drive the ideas in the story.

Bang on your keyboard and see the code that your actions generate. Drag your mouse around and watch how every movement is tracked by advertisers. Try your hand at debugging code or writing JavaScript. (I rocked the JavaScript.) And if you skim the article

—as I did just now, when I went back to choose my favorite interactive elements—expect to see a message like this at the end:

> Congratulations! You read 31558 words in 29 minutes, which is 1076 words per minute. Hahahahah as if. Nice. Cool. Frankly we expected no more of you.

Subversive. Animated. Hilarious. Not words you'd normally associate with a long-form technology piece, but there you have it. You will tell your friends about this article. You will become obnoxious at dinner parties, if you weren't already. You will go around feeling like more of a programming expert than you have any right to. Wasn't that fun?

I believe that entertainment value matters. I am the daughter of a working musician. Every Friday and Saturday night, and many weeknights as well, when gigs were plentiful, I watched my dad put on a tuxedo, splash a little Canoe on his chin, and walk out the door with a guitar case and amp. When he said good night to us, he used to say that he was "going out to entertain the folks."

That's not to say he doesn't take his music seriously. But he knows what his profession is about. He's an entertainer. His job is to go out and play for the folks.

And that's what our job is, as writers. Our job is to go and play for the folks.

Last summer I spent a week in Rhode Island with my aunt, who was dying of cancer. In case you didn't already know this, let me tell you that the palliative oncology ward of a hospital is about the worst place in the world. People were sobbing in the hallways. The "family room"—a sort of improvised retreat center equipped with televisions, couches, and vending machines—was almost always occupied by a woman crying into her cell phone. "You'd better get down here" was a common refrain heard in the family room. "We're losing her."

Horrible. And those were the people who *weren't* dying of cancer.

I got to know the ward pretty well, after spending five straight days there. Guess what I noticed as I walked the halls and spied on the other families who were going through the same tragedy I was?

Everyone had a book. Or a magazine. Or a newspaper. In the midst of all this suffering, on the very worst days of their lives, everyone was reading. And they weren't just reading escapist fiction. I saw history. Biography. Memoir. And—believe it or not—science and nature writing. I saw Rebecca Skloot, Diane Ackerman, and Wendell Berry on the oncology wing.

Even my aunt, in the last month of her life, was so relieved when I logged into her iPad and downloaded some audiobooks for her to listen to (she was nearly blind by then) during the long hours she spent alone, after we went home every night. She asked for an essay collection from Anne Lamott. "Not the one with all the cancer," she said, which, if you've read Anne Lamott, you know is not easy to find.

My point is this: everyone on that oncology wing needed something to read. They wanted an escape, sure, but they also wanted a connection. They wanted reassurance, hope, enlightenment, and understanding. We all read for the same reasons, just as we all listen to music for the same reasons.

Science writers get into the game with all kinds of noble, high-minded ambitions. We want to educate. To enlighten. To advocate. To accelerate the pace of progress. To win public support for research. To celebrate the work of incredibly intelligent and dedicated people who too often labor in obscurity.

But at the end of the day, we're all writers. We're just like novelists, memoirists, and poets. We're entertainers.

We're here to play for the folks.

Whether our audience is made up of congressional staffers or the miserable family members hunkered down in the hallways of the palliative-care wing, we are doing this for them.

The writers in this collection understand that. They're writing for us, and in many cases about us, in ways that make a real difference. Here are just a few examples:

The essay that had the most significant and immediate impact on people's lives was, in my opinion, Kathryn Schulz's "The Really Big One," about the risk of a catastrophic earthquake in the Pacific Northwest. Now, I might be biased, because the southern end of the Cascadia subduction zone burrows directly under my hometown of Eureka, California, which also happens to sit very near the San Andreas Fault. The last really big one, about 300 years ago, probably formed the bay I'm looking at right now.

I'm trying to come up with words to explain the impact that this article made on those of us who live in the Cascadia region, but it all ends up sounding like a bad pun in a newspaper headline: It sent shock waves through our community. It shook us up. It rocked our world. It caused a seismic shift in . . . well, you see the problem.

The point is that politicians, public safety agencies, and individual citizens throughout the Pacific Northwest changed their thinking because of this story. Some people changed their lives: I know people who moved away from the region entirely after the story ran, and I know people who bought and sold homes for the sole purpose of finding a safer place to live. Parents made contingency plans for their children's safety. Neighbors stocked up on emergency supplies. Contractors got busy bracing foundations.

On a larger scale, public officials have put a new emphasis on earthquake preparation. In June 2016, 6,000 emergency and military personnel participated in a four-day disaster-preparedness exercise across Oregon and Washington. A serious discussion is taking place right now among officials in all three West Coast states about implementing a regional early-warning system. In Portland, Oregon, the seismic safety of the city's critical bridges and its many unreinforced masonry buildings have become a top priority. And along the coast—including right here where I live—people in small towns are preparing for the real possibility of a tsunami evacuation.

I simply cannot overstate the power of this piece. When you read it, imagine that you live where I live. Your life would change because of this story, just like mine did. That's the power of great writing.

But this isn't the only piece that changed people's lives this past year. Sarah Maslin Nir's investigation into the health, safety, and working conditions of New York's nail-salon workers highlights the dangers faced by immigrant workers who toil in a highly visible yet often overlooked environment. The report, which was published not only in English but in Chinese, Korean, and Spanish, led to increased enforcement and a higher level of scrutiny by public health and employment officials. We've included the more health- and science-focused second half, "Perfect Nails, Poisoned Workers," here, but I urge you also to read the first half, about wages and hours, if you haven't already.

Gabrielle Glaser's "The False Gospel of Alcoholics Anonymous" is another terrific piece of scientific reporting that directly addresses the well-being of a marginalized group of people: alcoholics and drug addicts. While wealthier people might have the means to avail themselves of a wide range of treatment options, poor people and people whose addiction has landed them in jail often have few choices but to enter the treatment programs mandated by judges or social workers. And when that treatment program—the faith-based 12-step program—fails to stand up to scientific rigor, they aren't provided a better option.

By shining a light on those treatment methods that can be backed up by solid medical studies and those that cannot, Glaser's piece exposes the failures of our health care and criminal justice systems when it comes to treating addicts. (Her work might also benefit wealthier people seeking treatment: a month's inpatient treatment at a rehab center starts at $40,000 and includes such unproven treatment methods as art therapy and mindfulness mazes, while far more effective treatments cost only a few thousand dollars.)

Medical professionals in any other field follow clinical trials, monitor the results of scientifically valid studies, and update their treatment practices accordingly. But this doesn't happen nearly as often in the field of addiction treatment—and that's a huge disservice to the people whose very lives depend upon its effectiveness. Glaser's piece exposes the flaws in our treatment of addiction and will, I hope, bring about change.

Charles Mann's article "Solar, Eclipsed" takes a perspective on climate change that we don't hear enough in this country: that of the unwired poor. In India, carbon emissions continue to grow—and have to, if the 300 million Indians who currently live without electricity are to have any at all (which is to say nothing of the millions more who have only intermittent, unreliable power). As Mann points out, Westerners tout alternative energy as the solution to India's energy needs, but his visit to the remote village of Luckman shows how difficult this might be. Come nightfall, a single solar-powered 6-watt LED lamp is the only light source for the entire village. It's easy for Americans and Europeans to propose this solution for India's poor, but none of us would be satisfied with that result. The answers won't come easily, but this piece raises important questions.

Many of the stories I chose for this collection delighted me because they spoke to perspectives I don't hear often enough or don't understand. I want to look at climate change from the perspective of India's poorest. I want to do more to protect the safety of immigrant workers in nail salons. I want great science and nature writing to take me into worlds I can't imagine and to challenge my ideas about what matters.

But how do we decide what matters? This is the part of the essay where I need to talk about the underwear story. You might be wondering why Rose Eveleth's "Why Are Sports Bras So Terrible?" ended up here and not, say, in *Best American Fashion Writing*. It turns out that building a functional sports bra depends a great deal on engineering and technology. It also turns out that the existence of a functional sports bra directly affects women's health.

I'm particularly enamored of this essay because it taught me something very interesting about my own biases as a reader. A piece on the technology and psychology behind athletic wear should hit me right in the demographic: I've been wearing sports bras for 30 years, I care a great deal about fitness and women's access to sports, and I'm predisposed to champion a story about an issue particular to women that might otherwise be sidelined.

This story was well researched, it was interesting, and it made solid arguments. But the truth is, I kept arguing with it in my own mind because it didn't fit with my experience. *I've never had a problem with sports bras,* I found myself thinking. *Maybe this isn't really an issue.*

So I put out a call on social media and asked women friends to report in if sports bra–related problems ever prevented them from getting exercise or playing a sport. The responses were immediate and convincing. I heard from women who went years without getting any kind of exercise. Others gave up running, dancing, and most other types of cardio workouts. I heard from women who dropped out of sports, and others who suffered serious shoulder and neck pain from ill-fitting bras or from "double bagging," a practice I had never even imagined, which involves wearing one bra on top of another in an often futile attempt to get enough support.

I point this out because it shows how easily a worthy topic can be marginalized if the person in charge of making the selection—in this case, me, the editor of this collection—doesn't have a life experience that fits the story. Rose Eveleth made the case quite perfectly: "Breast discomfort is a leading reason women stop participating in sports. And in extreme cases, an ill-fitting bra can actually do nerve damage."

As much as I care about—and am personally involved in—fitness, sports bras, and women's issues, I kept asking myself whether this issue really mattered. Why? Because in my decades as a sports bra–wearing, cardio-loving feminist, I had never heard of it.

So. Lesson learned. We could all benefit from looking outside our own experiences once in a while. I specifically went looking for stories that help us do that.

Some of the pieces I enjoy most in this collection take a different perspective on the very idea of science and nature writing. So much of the best writing in the field is told from the perspective of a journalist trekking along behind a scientist and attempting to explain what exactly he or she does and why it matters. The writer is an enthusiastic observer who takes up the task of defining the problem (the encroachment of bark beetles, the intractability of certain cancers), building the world in which these problems exist (the Rocky Mountains, a public housing project) and then embodying for the reader the scientist who hopes to do something about it. It's the "hero's journey" narrative structure, superimposed upon science journalism.

We love these stories. We love the idea of the lone protagonist battling a diabolical and menacing enemy. I adore that form of storytelling and was happy to include many such pieces here. But I was also delighted to find writers who had a different kind of story to tell and an interesting structure to hang it on.

Katie Worth's "Telescope Wars" is one such piece. It tells the distressing history of rival astronomers competing to build the world's largest telescope. For 15 years—well, actually, for nearly 100 years, depending on how far back you want to go—astronomers have launched rival projects to build telescopes so powerful that they could gaze deep into the cosmos and bring back remarkably clear and detailed images, the likes of which we've never seen. But these supersized telescopes don't yet exist, and none of the competing projects have enough funding. It all comes down to an inability to collaborate that has been fed by, as Worth puts it, "personality conflicts, miscommunications, competing technologies, and an expanding universe of bitterness."

I love this piece because it calls attention to what isn't getting done. It looks at what scientists have failed to accomplish. It calls astronomers to task for failing to see beyond petty rivalries to work together on a project of global significance. It asks why there was never "a little adult supervision in the room." In other words, it looks honestly at the messy process of actually getting science done. That's a perspective we don't hear often enough.

I'm also pleased to include some serious pieces of investigative reporting, in which the author is far more than an observant bystander. Bryan Christy's "Tracking Ivory" is the firsthand account of an ambitious undertaking: to fabricate ivory tusks, implant GPS and satellite trackers within them, and release them into the clandestine ivory market, with the goal of finding out how the tusks of endangered elephants are trafficked around the world.

This is not a risk-free project. Christy spends the night in a Tanzanian jail and embeds himself with a heavily armed antipoaching patrol as he journeys to the heart of the illegal ivory trade. Ultimately, he has to part ways with his GPS-embedded artificial ivory and track its movements on a computer screen. It moves 600 miles into Sudan, illuminating on a map the route that smugglers go to such great lengths to keep secret. With over 30,000 African elephants killed every year, the stakes could not be higher. Christy and his team at *National Geographic* should be commended for this exemplary piece of journalism.

Finally, this collection celebrates language. You'll read some of the most lyrical and poetic writers exploring the themes of science and nature today. Chelsea Biondolillo's "Back to the Land" is a lovely short piece on a gruesome topic: a research facility in which forensic anthropologists study how bodies decay in the scorching Texas sun. I'm not going to quote from it here because I don't want to ruin it for you, but I promise those last few lines will take your breath away.

Gaurav Raj Telhan's "Begin Cutting," about the dissection of cadavers in medical school, is just as macabre but also gorgeous. "Her eyelids," he writes, "not entirely shut, revealed the green of her irises, afloat in a white scleral sea." Later, when he begins the dissection, he recounts how "in my beginner's hands, the tapping of the scalpel against the spine sounded like beats of Morse code."

Amazing. Dazzling.

There is also the beautiful and heartbreaking "My Periodic Table," published about a month before Oliver Sacks's death. I can think of very few writers who have brought more lyricism to science and nature writing than Sacks. Through his writing we knew his warmth, curiosity, hunger, and love of language. His work has appeared in this collection in 10 out of the 16 years it's been in publication, and it pains me to think that it won't appear here again.

Finally—Amy Leach, where have you been all my life? Her delightful short piece "The Modern Moose" sent me right out in search of everything else she's ever written. This brief and breathtaking tribute to the moose ("as modern as Mugellini and should be coequally respected") is so startling that I could do nothing but make a list of adjectives, which I present here in alphabetical order: brilliant, fantastical, imaginative, irreverent, reverent, subversive, surreal, tricky, unconventional, unworldly, weird, wonderful.

I could go on, but you get the idea.

Thanks to series editor Tim Folger, Houghton Mifflin Harcourt editor Naomi Gibbs, and also to former editors Mary Roach and Deborah Blum, who offered wisdom and encouragement along the way. Thanks to my husband, Scott Brown, for reading a hundred or so wonderful science and nature essays along with me this year and talking about them every night for many, many weeks. And thanks to all the writers who show up to entertain the folks. Here's to 2016.

AMY STEWART

The Best American Science and Nature Writing 2016

The Best American Science and Nature Writing 2016

CHELSEA BIONDOLILLO

Back to the Land

FROM *Orion*

NOT FAR FROM AUSTIN, a dirt road winds through patches of bluestem, spiked mesquite, and twisted live oaks on the working Freeman Ranch, owned by Texas State University. The ranch is home to an organic garden, a lively herd of cattle, and the Texas State Forensic Anthropology Research Facility, or FARF. It is also home to FARF's ongoing research project on scavenging and decomposition, which is led by two forensic anthropologists.

Another way of saying this is that if you travel for long enough on a dirt road just outside San Marcos, and if you are able to get through the double security gates, what you will find amid the grasses and trees and occasional longhorn is a field of human bodies in varying states of decay.

I'm here to ask one of the researchers about vultures. Specifically, I want to understand how the birds are helping law enforcement to develop crime-scene-investigation protocols. The researcher gives me a pair of blue hospital-style booties to slip over my shoes just inside the second gate.

The anthropologists are studying two things: how the Texas sun turns a body into a rusted mummy feeding the switchgrass, and how vultures scavenge those bodies. To learn about the former, the donated cadavers are laid under metal mesh cages. To understand the latter, they are left exposed, tagged wrists crossed, or open, as in savasana—the corpse pose.

Except nothing I see in the open looks *reposed.* The spines are twisted, the bones scattered. "This young man died in a violent accident," my guide says as she crouches down to point out signs of vultures, certain cracks and predictable breaks the birds leave behind on the bones. Families donate the bodies of loved ones for a variety of reasons, she says. So the deceased can continue to teach, for example.

My eyes stay wide open and my mouth stays mostly shut as we walk through the grass. I try to think of what I'm seeing as former people, but I can't. The people have left. All that remains are remains: a countable collection of bones. Shin is connected to leg; leg is connected to hip. I stare at teeth, my notebook full of questions forgotten. I am trying to place birds there, in that mouth, or this eye. But the birds, too, have left. They circle high above us; even their specks have vanished.

The work the researchers and their teams of student volunteers undertake is helping law enforcement in the borderlands solve crimes and identify the dead. That's why they place donated bodies in the fields, why they work first under the fierce sun, monitoring rates of decay and dispersal, and later in the lab, with toothbrushes and dish soap, scrubbing off stubborn shreds of tendon and cartilage. They need to know how long until the birds come, and how long before they leave, so that a coroner out in the desert has data he can look to when he makes his time-of-death estimate. How long has this body been here? How long has this person been lost?

I remember a story I'd read about butterflies scavenging ammonia salts from carrion. I ask the anthropologist if she's ever seen butterflies out here. I cannot say the word "carrion." Instead, I gesture toward the low cages when I say "out here."

"Oh my, yes," she says. Her voice is soft with a gentle Texas lilt. "In the spring, it's quite beautiful—all the black-eyed Susans are in bloom and there are butterflies all over . . . It's a sight."

For a moment, I forget the talons and curved beaks of vultures. I forget the frenzy of feathers. Instead, I imagine a field of migrating monarchs, green sulfurs, orange-eyed buckeyes, and gilt-edged mourning cloaks. This field becomes their wished-for respite during a long journey. The deliberate opening and closing of all those wings becomes a kind of breath, a last sigh that reaches at once down into the roots of the bluestem and up into the flyway, following the current all the way to Mexico.

BRYAN CHRISTY

Tracking Ivory

FROM *National Geographic*

WHEN THE AMERICAN MUSEUM of Natural History wanted to update the Hall of North American Mammals, taxidermist George Dante got the call. When the tortoise Lonesome George, emblem of the Galápagos Islands, died, it was Dante who was tasked with restoring him. But Dante, who is one of the world's most respected taxidermists, has never done what I'm asking him to do. No one has.

I want Dante to design an artificial elephant tusk that has the look and feel of confiscated tusks loaned to me by the U.S. Fish and Wildlife Service. Inside the fake tusk, I want him to embed a custom-made GPS and satellite-based tracking system. If he can do this, I'll ask him to make several more tusks. In the criminal world, ivory operates as currency, so in a way I'm asking Dante to print counterfeit money I can follow.

I will use his tusks to hunt the people who kill elephants and to learn what roads their ivory plunder follows, which ports it leaves, what ships it travels on, what cities and countries it transits, and where it ends up. Will artificial tusks planted in a central African country head east—or west—toward a coast with reliable transportation to Asian markets? Will they go north, the most violent ivory path on the African continent? Or will they go nowhere, discovered before they're moved and turned in by an honest person?

As we talk over my design needs, Dante's brown eyes sparkle like a boy's on Christmas morning. To test ivory, dealers will scratch a tusk with a knife or hold a lighter under it; ivory is a tooth and won't melt. My tusks will have to act like ivory. "And I gotta find a way to get that shine," Dante says, referring to the gloss a clean elephant tusk has.

"I need Schreger lines too, George," I say, referring to the cross-hatching on the butt of a sawn tusk that looks like growth rings of a tree trunk.

Like much of the world, George Dante knows that the African elephant is under siege. A booming Chinese middle class with an insatiable taste for ivory, crippling poverty in Africa, weak and corrupt law enforcement, and more ways than ever to kill an elephant have created a perfect storm. The result: some 30,000 African elephants are slaughtered every year, more than 100,000 between 2009 and 2012, and the pace of killing is not slowing. Most illegal ivory goes to China, where a pair of ivory chopsticks can bring more than a thousand dollars and carved tusks sell for hundreds of thousands of dollars.

East Africa is now ground zero for much of the poaching. In June the Tanzanian government announced that the country has lost 60 percent of its elephants in the past five years, down from 110,000 to fewer than 44,000. During the same period, neighboring Mozambique is reported to have lost 48 percent of its elephants. Locals, including poor villagers and unpaid park rangers, are killing elephants for cash—a risk they're willing to take because even if they're caught, the penalties are often negligible. But in central Africa, as I learned firsthand, something more sinister is driving the killing: militias and terrorist groups funded in part by ivory are poaching elephants, often outside their home countries, and even hiding inside national parks. They're looting communities, enslaving people, and killing park rangers who get in their way.

South Sudan. The Central African Republic (CAR). The Democratic Republic of the Congo (DRC). Sudan. Chad. Five of the world's least stable nations, as ranked by the Washington, D.C.–based organization the Fund for Peace, are home to people who travel to other countries to kill elephants. Year after year, the path to many of the biggest, most horrific elephant killings traces back to Sudan, which has no elephants left but gives comfort to foreign-born poacher-terrorists and is home to the Janjaweed and other Sudanese cross-continental marauders.

Park rangers are often the only forces going up against the killers. Outnumbered and ill-equipped, they're manning the frontline in a violent battle that affects us all.

Garamba's Victims

Garamba National Park, in the northeast corner of the DRC and on the border with South Sudan, is a UNESCO World Heritage site, internationally famous for its elephants and its boundless ocean of green. But when I ask a gathering of children and elders in the village of Kpaika, about 30 miles from the park's western border, how many of them have visited Garamba, no one raises a hand. When I ask, "How many of you have been kidnapped by the LRA?"—I understand why.

Father Ernest Sugule, who ministers to the village, tells me that many children in his diocese have seen family members killed by the Lord's Resistance Army, or LRA, the Ugandan rebel group led by Joseph Kony, one of Africa's most wanted terrorists. Sugule is the founder of a group that provides assistance to victims of Kony's army. "I've met more than a thousand children who have been abducted," he says as we talk inside his church in the nearby town of Dungu. "When they're abducted, they're very young, and they're forced to do horrible things. Most of these children are very, very traumatized when they come back home." They have nightmares, Sugule continues. They have flashbacks. Their own families are afraid that they're devils, or forever soldiers, who might kill them in the night. It is assumed that the girls were raped, so it's difficult for them to find husbands. Villagers sometimes taunt returned children with the same expression used for Kony's men: "LRA Tongo Tongo." "LRA Cut Cut"—a reference, Sugule explains, to the militants' vicious use of machetes.

Kony is a former Roman Catholic altar boy whose stated mission is to overthrow the Ugandan government on behalf of the Acholi people of northern Uganda, and to rule the country according to his version of the Ten Commandments. Since the 1980s, and beginning in Uganda, Kony's minions are alleged to have killed tens of thousands of people, slicing the lips, ears, and breasts off women, raping children and women, chopping off the feet of those caught riding bicycles, and kidnapping young boys to create an army of child soldiers who themselves grow into killers.

In 1994 Kony left Uganda and took his murderous gang on the road. He went first to Sudan, initiating a pattern of border hopping that continues to make him difficult to track. At the time Sudan's north and south were in a civil war, and Kony offered Sudan's government, in Khartoum, a way to destabilize the south. For 10 years Khartoum supplied him with food, medicine, and arms, including automatic rifles, antiaircraft guns, rocket-propelled grenades, and mortars. It was thanks largely to efforts by the group Invisible Children and its video *Kony 2012* that Kony became a household name in the West. In the United States, Presidents George W. Bush and Barack Obama supported efforts either to arrest or kill him. The U.S. State Department named Kony a "specially designated global terrorist" in 2008, and the African Union has designated the LRA a terrorist organization.

When north and south Sudan signed a peace agreement in 2005, Kony lost his Sudanese host. In March 2006 he fled for the DRC and set up camp in Garamba National Park, then home to some 4,000 elephants. From Garamba, Kony signaled his desire for peace with Uganda, sending emissaries to neutral Juba, in southern Sudan, to negotiate with Ugandan officials while he and his men lived unmolested in and around the park, protected by a cease-fire agreement. His army farmed vegetables. Kony even invited foreign press into his camp for interviews. Meanwhile, flouting the cease-fire, his men crossed into CAR, where they kidnapped hundreds of children and made sex slaves of women they brought back to the park.

Father Sugule introduces me to three young girls, recent LRA kidnapping victims, who are sitting on a wooden bench in his church. Geli Oh, 16, spent longer with Kony's army than her two friends—two and a half terrible years. She looks at the floor while her friends whisper to each other, smile radiantly, and nibble on cookies we've brought for them. Geli Oh perks up at the word "elephant." She saw many elephants in Garamba National Park, she says, which is where the LRA took her. Tongo Tongo shot two elephants one day, she says. "They say the more elephants they kill, the more ivory they get."

Kony's force has declined from a peak of 2,700 combatants in

1999 to an estimated 150 to 250 core fighters today. Killings of civilians have likewise dropped, from 1,252 in 2009 to 13 in 2014, but abductions are rising again, and it takes the arrival of only a few of the armed militants to send fear ricocheting through communities. In village after village along the road between Father Sugule's church and what is now South Sudan, I meet Kony victims who describe being fed elephant meat and how, after elephants were killed, militants took the ivory away.

But where?

The Problem Solver

To follow my artificial tusks from the jungle to their final destination, I need a tracking device capable of transmitting exact locations without dead zones. It needs to be durable and small enough to fit inside the cavities George Dante would make in the blocks of resin and lead that formed the tusks. Quintin Kermeen, 51, based in Concord, California, has the credentials, and the personality, I'm looking for. Kermeen started in the radio-tracking business when he was 15 and has since built electronic trackers and collars for wildlife from Andean bears to California condors to Tasmanian devils. He designed a GPS tracker that the U.S. Geological Survey embedded in live Burmese pythons to monitor the invasive snakes in the Florida Everglades. For his "Judas pig" project, he built GPS satellite collars to enable pest-control authorities in New Zealand to send feral pigs into the bush and locate their invasive piggy friends. We meet over Skype.

"You must be a real animal lover," I say.

"I'm not an animal lover," he snaps. "I'm a problem solver."

I laugh. "Then you're just the man for me."

After months of tinkering, Kermeen's final bespoke ivory-tracking device arrives in the mail. It consists of a battery capable of lasting more than a year, a GPS receiver, an Iridium satellite transceiver, and a temperature sensor.

While Dante set about embedding Kermeen's tracker inside his tusk mold, a third team member, John Flaig, a specialist in near-space, balloon-based photography—images taken from at least the height of spy planes—was preparing to monitor the tusks as they moved. Using Kermeen's technology, he could adjust how many times a day they tried to communicate with a satellite via the Internet. We would follow them using Google Earth.

"I Want Ivory for Ammunition"

On September 11, 2014, Michael Onen, a sergeant in Kony's army, walked out of Garamba National Park carrying an AK-47, five magazines of ammunition, and a story. Onen is short and looks even smaller wearing a camouflage-patterned Ugandan army uniform that's too long for him in the sleeves. He sits on a plastic chair opposite me in a clearing at the African Union forces base in Obo, in the southeastern corner of CAR, where he is in custody. Onen had been part of an LRA poaching operation in Garamba consisting of 41 fighters, including Kony's son Salim. The operation was designed by Kony himself, Onen says. During the summer Kony's soldiers had killed 25 elephants in Garamba, and they were on their way back to Kony carrying the ivory.

Around us stroll Ugandan army soldiers, who make up the entire African Union contingent based in Obo and are committed to finding and killing Kony. The soldiers embrace Onen as one of their own, and in fundamental ways he is. He was 22 years old the night in 1998 that Kony's soldiers raided his village in Gulu, Uganda, and pulled him from his bed. His wife, abducted later, was killed.

From the moment of his capture, Onen says, he was a complainer. Being small, he balked at having to carry the heavy bundles that Kony's militants ferry from camp to camp in their patrols across central Africa, and for his whining, he was beaten with a machete. But Onen got his way. Instead of being made a soldier, he was designated a signaler—a radioman privy to Kony's secret communications.

During the failed peace talks with Uganda, while Kony hid in Garamba from 2006 to 2008, Onen had been assigned to Kony's lead peace negotiator, Vincent Otti. Otti liked elephants, Onen recalled, and forbade their killing. But after Otti left Garamba to participate in the peace talks, Kony began killing elephants for ivory.

Otti was furious, Onen says. "Why are you collecting ivory?" Otti demanded of Kony. "Aren't you interested in peace talks?"

No, I want ivory for ammunition to keep fighting, was Kony's reply, according to Onen, who was listening to transmissions. "Ivory operates as a savings account for Kony," says Marty Regan, of the U.S. State Department's Bureau of Conflict and Stabilization Operations. Kony's army had arrived in Garamba in 2006 with little ammunition left to continue its war, Onen tells me. "It's only the ivory that will make the LRA strong," he recalls Kony saying.

Instead of signing a peace agreement, Kony had his peace negotiator executed.

From Garamba, Kony sent an exploratory team to Darfur to look into forging a new relationship with the Sudan Armed Forces (SAF), who had supported him against Uganda, hoping to exchange ivory for rocket-propelled grenades and other weapons. Meanwhile, according to Onen, Kony's men hid ivory by burying it in the ground or submerging it in rivers. His account was corroborated by Caesar Achellam, a former intelligence chief for Kony who is now in the Ugandan government's custody. Achellam told me that Kony's men planned for the future. He said they bury sealed buckets of water along parched travel routes and bury ivory for safekeeping as well.

"They can get what they want today," he said, "and keep it there for two, three, or even more than five years."

The Ugandan military finally attacked Kony's Garamba camps in late 2008. The airstrike, dubbed Operation Lightning Thunder, included support from the DRC, southern Sudan, and the United States. But it failed to rout Kony or his leadership. Kony's response was immediate and savage. Beginning on Christmas Eve, his soldiers spread out in small teams and murdered civilians. In three weeks Kony's brutes killed more than 800 people and kidnapped more than 160 children. The UN estimated that the massacre displaced more than 100,000 Congolese and Sudanese. On January 2, 2009, the horror bled into Garamba's headquarters, at Nagero, where Kony's soldiers burned the park rangers' main building, destroyed equipment, and killed at least 8 rangers and staff members.

Six years later, on October 25, 2014, Onen tells me, his poaching mission to Garamba was scheduled to deliver its ivory to Kony in Sudan. Kony was adamant in his radio transmissions. "Do not lose even one tusk," he instructed the group, according to Onen, who said the plan was to carry the ivory to a rendezvous in CAR and then on to a market town in Darfur called Songo, not far from the Sudan Armed Forces garrison in Dafaq. There, Onen says, Kony's men trade ivory with the Sudanese military for salt, sugar, and arms. The relationship is close: "SAF warns Kony if there's trouble," Onen says.

As far as Onen knew, the poaching squad he abandoned was still making its way north from Garamba through CAR to Sudan. To me, it seems reasonable to think that the radioman's defection might have slowed the progress of the 25 elephants' tusks headed to Kony.

Maybe I could get my fake tusks to Kony too.

"You Are a Liar!"

An official in Dar es Salaam's international airport, in Tanzania —one of several countries I scouted for launching my tusks into the illegal trade—squints at an x-ray screen as my luggage rolls through his scanner.

"Open that one," he orders.

I unzip my suitcase to expose two fake tusks and hand him letters from the U.S. Fish and Wildlife Service and *National Geographic* certifying that they're artificial. A crowd gathers. Officials are pointing fingers and arguing. Those looking at the tusks think I'm an ivory trafficker. Those looking at the x-ray screen, which shows the trackers inside, think I'm smuggling a bomb. After more than an hour of animated debate, they phone the airport's wildlife expert. When he shows up, he picks up a tusk and runs his finger over the butt end. "Schreger lines," he says.

"Exactly," I say. "I had them . . ."

He points a finger at me and yells, "You are a liar, *bwana!*" (*Bwana* is Swahili for "sir.")

In 10 years he's never made a mistake, he says: the tusks are real. I spend a night in police custody, where I'm given a desk to sleep on. National Geographic television producer J. J. Kelley takes the floor in the waiting area. He asks for water for me and is led out of the building. When he returns hours later, he has three chicken dinners and several bottles of beer, paid for by the police chief. The three of us eat together (the police chief, a Muslim, leaves the beer to us). In the morning, after officials from Tanzania's Wildlife Division and the U.S. Embassy arrive, I'm released.

Our airport incident was one of many hiccups with the artificial tusks. Several Tanzanian officers who had presided over my arrest at the airport, including the wildlife expert, returned the next day to wish us bon voyage. "You did exactly what you were supposed to do," I said, shaking their hands.

It was reassuring to find the Tanzanian law enforcers so vigilant, because the country is plagued by perhaps the worst elephant poaching in Africa, and corruption is rife. In 2013 Khamis Kagasheki, then Tanzania's minister of natural resources and tourism, declared that the illegal ivory trade "involves rich people and politicians who have formed a very sophisticated network," and he accused four members of Tanzania's Parliament of being involved in it.

Garamba's Warriors

All around me I hear the click-clack of automatic weapons being loaded. I've flown from Garamba park headquarters to a dirt airstrip deep inside the park to join an antipoaching patrol. I arrive at what amounts to the park rangers' northern front, an outpost vulnerable both to Sudanese poachers and Kony's army. Here a ranger unit is permanently deployed to protect one of the park's most important assets: a radio tower that was being built. Garamba is managed through a partnership between the DRC's wildlife department and African Parks, a group based in Johannesburg, South Africa.

Since the 2008–9 attack by Kony's soldiers, rangers have finished building a new headquarters and acquired two airplanes and a helicopter. But ammunition is in perilously short supply—not even enough for basic training—and the rangers' largest weapon, a belt-fed machine gun, tends to jam every third round or so. The rangers I'm going out with have each been allocated a handful of rounds for old and unreliable AK-47s, most of them seized from poachers.

We plunge eight hours through elephant grass so tall and thick it's possible to get lost just 20 feet from the man in front of you—down grass ravines, up hills exposed to the enemy, across a murky, waist-deep pond. At the sound of a twig cracking or the detection of an unexpected scent on the wind, a ranger in front of me, Agoyo Mbikoyo, signals caution, and I drop with the team into a collective crouch and wait silently. I make a mental note that Kony's soldiers and other armed groups walk hundreds of miles from Sudan into this endless grass curtain to kill elephants. I wonder if Kony's men are out there now.

The recent death toll of elephants in Garamba has been staggering, even by central African standards. Poachers killed at least 132 last year, and as of this June, rangers had discovered another 42 carcasses with bullet holes, more than 30 attributed to a single Sudanese poaching expedition—a combined loss amounting to more than 10 percent of the park's entire population of elephants, estimated now to be no more than about 1,500.

From March 2014 to March 2015 Garamba's rangers recorded 31 contacts with armed poachers, more than half of whom were with groups traveling south from the direction of South Sudan and Sudan. They included South Sudanese armed forces (SPLA) and Sudanese military, as well as defectors from those militaries and an assortment of Sudan-based rebels. Congo's own soldiers threaten the park's southern border, and villagers around the park sometimes poach elephants too. And someone—it's unclear who—is believed to be killing elephants from helicopters, as evidenced by bullet holes in the tops of skulls and the removal of tusks by what can only be chainsaws.

"My interpretation," says Jean Marc Froment, then director of the park, is that the Ugandan military "is conducting operations inside Garamba and at the same time taking some ivory." But, he adds, the poachers could be SPLA, which uses the same type of helicopter seen over the park. An adviser to the Ugandan military rejects the helicopter accusation, and suggests that the elephants might have been shot in the top of the head after they were down.

Having worked extensively throughout central Africa, Froment transferred to Garamba in early 2014 after rangers discovered dozens of elephant carcasses in the park. It was supposed to be a short posting, but he saw too much death to leave. He'd grown up not far from Garamba at a time when it was possible to fly over the park and see 5,000 elephants in a single gathering. Now it was rare to see 250 in a herd.

Froment uses the word "war" to describe the fight Garamba's 150 rangers are in with poachers. Money is available to outfit the rangers with better equipment, but buying new weapons requires formal approval of the Congolese army, something Froment has been unable to get.

Halfway through our patrol, we come upon a clearing of burned grass beside the Kassi River, the site of a recent battle between Garamba rangers and SPLA poachers, in which, rangers tell me, they killed two poachers. I find a human skull fragment, and I nearly pick up a live hand grenade near where the SPLA had camped, mistaking it for a baby tortoise. It hadn't exploded—yet.

All of central Africa is a hand grenade, its pin pulled by a history of resource exploitation from abroad, dictatorships, and poverty. "The poaching issue is a governance issue," Froment says. "We protect the elephant to protect the park. We protect the park to give the people something of value." He fights for elephants because he knows that without the animals' presence, no one will support Garamba, and the park—which he calls "Africa's heart"—will be lost. Garamba is a crucible within a crucible, a park under siege in a country often in civil war in a region that has nearly forgotten peace.

On our patrol we don't encounter any poachers or rebel groups. But time is stalking our team: months later, on April 25, 2015, while on patrol, the ranger who led me into Garamba, Agoyo Mbikoyo, was shot and killed by a gang of poachers. In June three more Garamba-based officers were killed. The culprits are believed to have been South Sudanese, according to African Parks.

Planting the Fake Tusks

After visiting Garamba, I arrange with a confidential source to put my tusks into the black market near Mboki, a small village in CAR midway between Garamba and Sudan that has been the target of attacks by Kony's army and where some people who have escaped from Kony have found safety. According to data stored in a GPS unit taken off the body of LRA commander Vincent "Binany"

Okumu, who was killed in a 2013 firefight with African Union forces on his return from poaching in Garamba, this village is on the path of ivory headed to Kony's base in Darfur.

Unwitting Targets

It was just after 4:00 a.m. on Heban hill, in Chad, 80 miles from the Sudanese border and 60 miles northeast of Zakouma National Park, home to the country's largest remaining elephant herd, 450 animals. Six antipoaching rangers and their cook, the entirety of the Hippotrague (French for "roan antelope") unit, were awake, dressed in camouflage uniforms, and preparing for morning prayers—devoted even in the darkness. It was the rainy season, and the rangers, like the elephants they were guarding, had left the park for higher ground.

Zakouma breathes its elephants. Dry season in, rainy season out. During the rains the park is more lake than land, and elephants split into two groups to escape the floods. One moves north toward Heban, the other west toward central Chad.

The rangers on Heban hill had little reason to be concerned for their safety. They were relieving a ranger team that had raided a Sudanese poachers' camp three weeks before and seized more than 1,000 rounds of ammunition; mobile phones holding photographs of bloated, dead elephants; a satellite phone with a solar-panel charger; two elephant tusks; a pair of camouflage pants; and a uniform with the insignia of Abu Tira—Sudan's notorious Central Reserve Police, alleged to have committed mass killings, assaults, and rapes in Darfur. The rangers also recovered a stamped Sudanese army leave slip granting three soldiers permission to travel from Darfur to a town near the Chadian border.

Zakouma National Park has lost nearly 90 percent of its elephants since 2002. Most—up to 3,000—were poached from 2005 to 2008. During those years Sudanese poachers arrived in groups of more than a dozen armed men, camping inside the park for months at a time, killing, in one instance, 64 elephants in a single hunt. When in 2008 the Wildlife Conservation Society introduced a surveillance airplane, poaching declined, but Sudanese marauders adapted, returning in hit squads of under six men, who infiltrated from outside the park on one-day hunts. They killed fewer elephants per hunt but were much harder to track and stop. Now, says the park's director, Rian Labuschagne, of African Parks, "my biggest fear is that they'll start coming in pairs."

The men of the Hippotrague unit assumed that after the previous team's raid, the poachers had all fled home. But instead, that morning the poachers were hiding among trees surrounding the rangers' camp. The poachers opened fire, killing five rangers. A sixth, a young lookout, ran down the hill, disappeared, and is presumed dead. The team's cook, also wounded, struggled 11 miles to get help. Later, when Labuschagne examined the trajectory of bullets at the scene, he concluded that the poachers had been trained in how to set up a crossfire, which, combined with evidence found at the scene, pointed to President Omar al-Bashir's Sudan Armed Forces.

The story typically would have ended with the wanton killing of these park rangers protecting elephants. But one of the murdered men, Idriss Adoum, had a younger brother, Saleh, who resolved that, when the rains stopped, he and a cousin would hunt the killers in Sudan, where so many ivory roads lead.

Sudan's Complicity

As Somalia is to piracy, Sudan has become to elephant poaching. In 2012 as many as 100 Sudanese and Chadian poachers on horseback rode across central Africa into Cameroon's Bouba Ndjida National Park. They set up camp and in a four-month rampage killed up to 650 elephants. According to Céline Sissler-Bienvenu, Francophone Africa director for the International Fund for Animal Welfare, who led a group into the park after the slaughter, the poachers were most likely from Darfur's Rizeigat tribal group, with ties to the Janjaweed—the violent, Sudanese-government-backed militias that have committed atrocities in Darfur. Sudanese and Chadian poachers were likewise implicated in the 2013 butchering of nearly 90 elephants—including 33 pregnant females as well as newborn calves—near Tikem, Chad, not far from Bouba Ndjida.

That members of the Sudanese military trade arms for ivory with the LRA raises questions about the highest levels of Sudan's government. In 2009 Bashir became the world's first sitting head of state indicted by the International Criminal Court (ICC) in The Hague for war crimes and crimes against humanity. In presenting that case, ICC prosecutor Luis Moreno-Ocampo underscored Bashir's control of the groups said to be behind Sudan's ivory trafficking: "He used the army, he enrolled the Militia/Janjaweed. They all report to him, they all obey him. His control is absolute."

Michael Onen, the defector from Kony's army, told me that the LRA and the Janjaweed had battled over ivory, with one group robbing the other, and that it was the Janjaweed's success in trading ivory that originally gave Kony the idea to start killing elephants. The LRA sells to the Sudan Armed Forces, Onen said.

Despite Sudan's role as a safe haven for groups known to traffic ivory, such as the LRA, Janjaweed, and other poaching gangs, the country has drawn limited official attention as a poaching state. The Convention on International Trade in Endangered Species of Wild Fauna and Flora (CITES), a treaty organization that governs international trade in ivory—and its continuing ban—has identified eight countries "of primary concern" when it comes to international ivory trafficking: China, Kenya, Malaysia, the Philippines, Thailand, Uganda, Tanzania, and Vietnam. Eight more are considered of secondary concern: Cameroon, Congo, the DRC, Egypt, Ethiopia, Gabon, Mozambique, and Nigeria. Three more are classified as of "importance to watch": Angola, Cambodia, and Laos.

Sudan is not on these lists, even though Sudanese poachers are a primary reason elephants are killed in several of the countries listed by CITES as of primary or secondary concern. Sudan is also a well-documented supplier of ivory to Egypt and is the recipient of substantial Chinese infrastructure investment, which typically comes with Chinese workers, a source of ivory smuggling in many parts of Africa. Ivory shops in Khartoum advertise in English and Chinese as well as Arabic. According to CITES secretary-general John Scanlon, Sudan does not appear on these lists because CITES sets priorities based mainly on ivory seizures, and there have been few ivory seizures linked to Sudan in recent years. Which raises the question: If ivory is poached by Sudanese, where is it going?

A Kony Hideout

My artificial tusks sit motionless for several weeks, a pair of tear-shaped blue dots on my computer screen, which displays a digital map of the eastern corner of CAR. Then, like a bobber in a fishing hole, a nibble. They shift a few miles. Suddenly they move steadily north, about 12 miles a day along the border with South Sudan, avoiding all roads. On the 15th day after they began to move, they cross into South Sudan and from there make their way into the Kafia Kingi enclave, a disputed territory in Darfur controlled by Sudan.

Kafia Kingi is so widely recognized as a Kony hideout that in April 2013 a coalition of groups, including Invisible Children, the Enough Project, and the Resolve, issued a report called *Hidden in Plain Sight: Sudan's Harboring of the LRA in the Kafia Kingi Enclave, 2009–2013*. LRA defectors I spoke with consistently placed the warlord in the Kafia Kingi area too. So did the African Union military forces, whose CAR-based men in Obo are tasked with finding Kony. "It's not a secret to anyone that Kony's in Sudan," says the State Department's Marty Regan. "It's his sanctuary."

A few days later the tusks proceed to Songo, the Sudanese market town where Onen said Kony's men trade ivory. In Songo the tusks are held for three days in what looks like a clearing outside town. Then they head south six miles, back into Kafia Kingi. I order a satellite shot of their location from DigitalGlobe, a commercial vendor of space imagery, and ask for outside help interpreting it. According to Colonel Mike Kabango, of the African Union forces, the image shows a large tent and two smaller ones; to Ryan Stage, a remote-sensing specialist in Colorado, it shows a large truck and two small tents. After three weeks the tusks turn north again, back into Sudan. Gathering speed, they continue north before abruptly turning east, in the direction of Khartoum.

Other roads also lead to Sudan. The relatives of murdered Zakouma ranger Idriss Adoum tracked one of the alleged Heban hill poachers to Sudan and arranged to have him brought back to Chad to stand trial. Soumaine Abdoulaye Issa had been in Darfur, he told a team of African Parks investigators, when he heard about an elephant-poaching mission to Chad led by a member of the Sudan Armed Forces. Issa, who is Chadian, said he joined the team of three Sudanese men and that together they rode more than two weeks to get to Heban, where they killed nine elephants in four days. After Zakouma's rangers destroyed their camp and confiscated their equipment, the poachers were unable to return to Sudan, so three weeks later they went back to Heban hill and attacked the Hippotrague unit.

Issa claimed he was merely a lookout, not a poacher. He wasn't contrite. In a public square in Am Timan, shortly before his trial, he shouted, "I know who betrayed me! I will escape from your jail, and I will kill him." He did escape, and a rumor in Zakouma is that he fled south to CAR.

"We've heard he went to Seleka," Idriss Adoum's son Issa tells me, referring to the violent rebel coalition that overthrew the CAR government on March 24, 2013. If true, Soumaine Issa will find poachers working with Seleka. Seleka and its rival, anti-Balaka, have set fire to people, thrown them off bridges, and murdered people wantonly, turning CAR into a lawless state—the kind of place where Kony's group and other terrorist organizations thrive. In May 2013 Seleka-backed Sudanese poachers attacked Dzanga Bai, an elephant oasis in Dzanga-Ndoki National Park of southwest CAR, killing 26 elephants. Dzanga Bai—also known as the village of elephants—is a mineral-rich mudhole where the animals congregate.

Earlier this year Kony suffered the defection of his commander of operations, Dominic Ongwen, who told African Union forces that Kony's desire for ivory was reinforced by Seleka. "Seleka rebels had a stock of about 300 ivory tusks that they sold, which enabled them to get the supplies that helped them overthrow President François Bozizé in CAR," Ongwen told African Union forces, according to his debriefing. Ongwen said Kony's plan is to obtain as much ivory as possible "for his future survival should he not be able to overthrow the government of Uganda."

Ongwen also said that Kony intends to form a squad to establish contact with Boko Haram, the Nigerian terrorist group responsible for widespread killings and the kidnappings of hundreds of Nigerian women and schoolgirls. Boko Haram also uses the bush as a base—Nigeria's Sambisa Forest, a game reserve south of Lake Chad. In March 2015 Boko Haram's leader, Abubakar Shekau, pledged allegiance to ISIS, and his group was renamed Islamic State's West Africa Province, giving that Middle East terrorist group a foothold in West Africa.

Where Next?

As of this writing, my artificial tusks sent out their last communication from a Sudanese town called Ed Daein, 500 miles southwest of

Khartoum. I know which house they're in: Using Google Earth, I see its light-blue roof on my screen. They're in a place 2.2 degrees Fahrenheit cooler than the ambient temperature, so perhaps they've been buried in the backyard. So far they've traveled 600 miles from jungle to desert in just under two months. Their path is consistent with the route Kony's defectors tell me ivory takes on the way to the warlord's Kafia Kingi base. By the time you read this, my tusks might have gone to Khartoum. Or possibly even shown up in illegal ivory's biggest consuming country: China.

Meanwhile, as leaders in Europe, the Middle East, and the United States strategize about how to stop the ever-expanding network of international terrorist organizations, somewhere in Africa a park ranger stands his post, holding an AK-47 and a handful of bullets, manning the frontline for all of us.

HELENE COOPER

They Helped Erase Ebola in Liberia. Now Liberia Is Erasing Them.

FROM *The New York Times*

IT WAS AROUND 3:00 in the afternoon when Sherdrick Koffa spotted, in neatly written script, the name on the body bag that he was preparing to set ablaze.

It was the name of a classmate. The two grew up together, had played together as children. Now, only a few days into his job burning the Ebola dead, work that had already estranged Mr. Koffa from his family, he was expected to burn the body of his friend.

He did it. First he sprayed the body with oil to help it catch fire. Then he carefully laid the body, along with several others, upon the kindling on the altar of the crematory. He stacked more kindling on top. Finally, as the kindling was lit with a torch, Mr. Koffa stripped off his protective gear and stalked off the field, away from the acrid smell of burning flesh.

He did not stop walking until he got home, and once there, he opened first one bottle, then two, of cane juice, the highly potent Liberian equivalent of moonshine. He drank all night, until he passed out.

Fifteen months later, Mr. Koffa is still drinking heavily.

It has been more than a year since this deeply religious country embraced one of its biggest taboos—cremating bodies—to rein in a rampaging Ebola pandemic. In that time, the majority of Liberians have started to move on.

But such is not the case for some 30 young men who were called upon during the height of the crisis last year.

As bodies were piling up in the streets and global health officials were warning that the country's ages-old traditions for funerals and burials were spreading the disease, these men did what few Liberians had done before: set fire to the dead. And for four months they did so repeatedly, burning close to 2,000 bodies.

Villagers protested near the site, hurling abuse and epithets at the men they called "those Ebola burners them." The government deployed police officers and soldiers along the dirt road to the crematory site in a field to keep angry locals from the men.

Their families shunned them as they pursued their grim work. One young man—Matthew Harmon—who lived not far from the crematory site here, said his mother refused to see him, telling him never to call again.

"My ma said, 'You burning body? Then I'nt want see you no more around me,'" Mr. Harmon said.

The ostracism darkened what was already an abysmal time for the men, so much so that now, a full year after the country has ceased the cremations, their lives remain virtually destroyed.

Their nights are spent with alcohol or drugs—habits they said they acquired to get through the mass burnings. One burner, William Togbah, says no night goes by when he does not dream of seared flesh. Several of the men, shunted aside by friends and family, now live together, sharing the same room in a house not far from the crematory site.

"I'm not in a good life now," Mr. Togbah said.

For the most part, Liberia has come out of its long national nightmare. Ebola cases flare up sporadically, with three new infections reported just last month, and experts warn that the disease may continue to pop up for years to come.

But children are back in school, crowding sidewalks in their uniforms as they head home in the afternoon. Soccer games have resumed, with a packed Antoinette Tubman Stadium recently hosting 10,000 people to watch their beloved Lone Star national team take on, and lose to, the African giants called Ivory Coast. Church pews have filled again, with people grasping one another's hands and trading hugs during the "peace be with you" part of services, a stark change from the no-touching rule many adopted here as the epidemic raged.

Yet the men continue to be tormented by what they saw and did. Initially they used an incinerator to burn the bodies, usually during the night. But that method left human bones to greet them when they returned in the morning, grisly remnants of the vibrant people who had lived their lives in this West African country.

Mr. Togbah and several others kept using the word "erase," as in, they erased the traces of the Ebola dead for their country. In turn, their country has now erased these young men.

Many Liberians still blame them for burning the dead. While they received certificates of appreciation from the Health Ministry, they were not part of the recognition ceremony held by the president to thank health-care workers for their efforts during the outbreak, an omission the young men took to heart.

"We missed some people," President Ellen Johnson Sirleaf said in an interview, adding that there were too many people to thank, and that she hoped to hold another event recognizing these men.

Still, they are largely shunned by Liberian society.

To understand how cremation is viewed by Liberians, one must first consider that this is a country with a national holiday—Decoration Day—meant solely for people to go and clean the graves of their loved ones. Every year on Decoration Day, Liberians troop to cemeteries and burial plots across the country with brooms, bleach, soap, and water.

Wakes can go on for days. People with little or no money to spare will beg and borrow to lay their dead in coffins made of black mahogany wood. They will build marble tombstones, and buy entire plots of land just to bury those they love. Many Liberians believe that if the dead are not properly buried, they will come back to haunt the living.

People here wash bodies and dress them to make sure they are ushered into the afterlife in style. A dead body for many Liberians is, in a sense, still a living thing, to be nurtured, looked after, and lovingly sent onward.

"It is just not in our culture to burn people," said Sampson Sayway, who helped organize the group of men to burn the bodies.

So when a line of government cars showed up in Marshall last year at the Indian-run field that is the country's sole crematory—previously used only for burning dead Indian nationals—Mr. Sayway, who lives a stone's throw away, immediately went out to investigate. It was early August 2014, at the height of the epidemic, and Ms. Johnson Sirleaf's besieged government had made a last-minute decision to take the advice of global health experts who said the bodies of the Ebola dead—the most infectious carriers of the disease—had to be burned.

Liberian officials knew the public would revolt. The government stationed police officers and soldiers along the route to keep villagers away. Government officials negotiated with Mr. Sayway over what the workers would be paid, around $250 a week. In a poor country like Liberia, that was enough money to get roughly 30 young men for the job.

But "it was no easy thing," said Fredrick Roberts, one of the burners, recalling that first night when the trucks came with the first 12 bodies. Terrified of getting too close to the Ebola dead, everyone scattered into the bush at first, as someone in the truck yelled out via a megaphone to keep a distance.

"I had no clue what I was getting into," said Ciata Bishop, who was tasked by the president with setting up the crematory operation.

That first night, the young men wore blue cloth jumpers and plastic gloves, but government officials later gave them protective clothing, gloves, and boots. Day after day, night after night, the trucks came with the bodies. The burners unloaded them, sprayed them with oil, and piled them on an altar.

"It smelled very bad," Mr. Koffa said. "Like meat, except different." His voice caught and he stopped talking, overcome. He and the other burners had gathered near the crematory field. They are never far away from it now. The place they hated so much has become a home, of sorts. Nowhere else will accept them.

"They would bring us thirty, sixty, one hundred bodies a day," Mr. Koffa said.

Because the incinerator was unable to turn the bones to ash, the men switched to burning bodies on pyres set upon two altars in the field. It was more time-consuming, but at least at the end there were only ashes to deal with.

One day the trucks delivered 137 bodies. "It took two days and a half," said Burdgess Willie, another burner. "It smelled so bad, we kept having to go away and then come back."

Sometimes there were explosions, from the combustion of the oil, body bags, and wood. The noise terrified villagers, further adding to their anger at the burners and the process.

Mr. Roberts's landlord put him out of his rented room, and he moved in with Mr. Harmon, the burner whose mother had shunned him. Soon other young men, turned out of their homes, were sharing the small room, too.

The men took to drinking and drugs to get through the nights. Government officials sent them extra bottles of cane juice, they said.

"When you see fifty, seventy, eighty bodies like that every day, that the only way you can make it," Mr. Willie said.

Then suddenly, just like that, it was over. In December, under intense public pressure and with the number of Ebola deaths declining, the government announced that it was ending cremations. A new 25-acre parcel had been secured, government officials said, to bury the Ebola dead.

For the 30 young men who carried out the task of burning more than 2,000 Ebola dead, the ordeal was over.

Except it wasn't. "People still mock at us," Mr. Roberts said. "When they see us, they say, 'That's Ebola burner them, oh.'"

Through the ordeal, the young men said they thought they would get government scholarships when it was all over. They thought they would be hailed as heroes, that people would apologizc for shunning them. They are still waiting.

Mr. Roberts said that a few days ago, almost a year since the government ended the cremations, he tried to get into a taxi. One of the passengers spotted him and quickly turned to the driver. "He said, 'This man worked in the fence, that Ebola burner, oh,'" Mr. Roberts recalled.

The response came quickly.

"Get down from the car," the taxi driver insisted.

GRETEL EHRLICH

Rotten Ice

FROM *Harper's Magazine*

I FIRST WENT TO Greenland in 1993 to get above tree line. I'd been hit by lightning and was back on my feet after a long two-year recovery. Feeling claustrophobic, I needed to see horizon lines, and off I went with no real idea of where I was going. A chance meeting with a couple from west Greenland drew me north for a summer and part of the next dark winter. When I returned the following spring, the ice had failed to come in. I had planned to travel up the west coast by dogsled on the route that Knud Rasmussen took during his 1916–18 expedition. I didn't know then that such a trip was no longer possible, that the ice on which Arctic people and animals had relied for thousands of years would soon be nearly gone.

In the following years I went much farther up the coast, to the two oldest northernmost villages in the world: Qaanaaq and Siorapaluk. From there I traveled with an extended family of Inuit subsistence hunters who represent an ice-evolved culture that stretches across the Polar North. Here, snowmobiles are banned for hunting purposes; against all odds, traditional practices are still carried on: hunting seals and walrus from dogsleds in winter, spring, and fall; catching narwhals from kayaks in summer; making and wearing polar-bear pants, fox anoraks, sealskin mittens and boots. In Qaanaaq's large communal workshop, 21st-century tools are used to make Ice Age equipment: harpoons, dogsleds, kayaks. The ways in which these Greenlanders get their food are not much different than they were a thousand years ago, but in recent years Arctic scientists have labeled Greenland's seasonal sea ice "a rotten ice regime." Instead of nine months of good ice, there are only two or three. Where the ice in spring was once routinely 6 to 10 feet thick, in 2004 the thickness was only 7 inches even when the temperature was –30 degrees Fahrenheit. "It is breaking up from beneath," one hunter explained, "because of the wind and stormy waters. We never had that before. It was always clear skies, cold weather, calm seas. We see the ice not wanting to come back. If the ice goes it will be a disaster. Without ice we are nothing."

Icebergs originate from glaciers; ice sheets are distinct from sea ice, but they, too, are affected by the global furnace: 2014 was the hottest year on earth since record keeping began, in 1880. Greenland's ice sheet is now shedding ice five times faster than it did in the 1990s, causing ice to flow down canyons and cliffs at alarming speeds. In 2010 the Petermann Glacier, in Greenland's far north, calved a 100-square-mile "ice island," and in 2012 the glacier lost a chunk twice the size of Manhattan. Straits and bays between northwest Greenland and Ellesmere Island, part of Canada's Nunavut territory, are often clogged with rotting, or unstable, ice. In the summer of 2012 almost the whole surface of Greenland's ice sheet turned to slush.

What happens at the top of the world affects all of us. The Arctic is the earth's natural air conditioner. Ice and snow radiate 80 percent of the sun's heat back into space, keeping the middle latitudes temperate. Dark, open oceans and bare land are heat sinks; open water eats ice. Deep regions of the Pacific Ocean have heated 15 times faster over the past 60 years than during warming periods in the preceding 10,000, and the effect on both glaciers and sea ice is obvious: as warm seawater pushes far north, seasonal sea ice disintegrates, causing the floating tongues of outlet glaciers to wear thin and snap off.

By 2004 the sea ice in north Greenland was too precarious for us to travel any distance north, south, or west from Qaanaaq. Sea ice is a Greenlander's highway and the platform on which marine mammals—including walrus, ring seals, bearded seals, and polar bears—Arctic foxes, and seabirds travel, rest, breed, and hunt. "Those times we went out to Kiatak and Herbert Islands, up Politiken's Glacier, or way north to Etah and Humboldt Glacier," the Inuit hunters said, "we cannot go there anymore." In 2012 the Arctic Ocean's sea ice shrank to a record minimum. Last year the rate of ice loss in July averaged 40,000 square miles per day.

The Greenland ice sheet is 1,500 miles long, 680 miles wide, and covers most of the island. The sheet contains roughly 8 percent of the world's freshwater. GRACE (Gravity Recovery and Climate Experiment), a satellite launched in 2002, is one of the tools used by scientists to understand the accelerated melting of the ice sheet. GRACE monitors monthly changes in the ice sheet's total mass, and has revealed a drastic decrease. Scientists who study the Arctic's sensitivity to weather and climate now question its stability. "Global warming has fundamentally altered the background conditions that give rise to all weather," Kevin Trenberth, a scientist at the National Center for Atmospheric Research, in Boulder, Colorado, says. Alun Hubbard, a Welsh glaciologist, reports: "The melt is going off the scale! The rate of retreat is unprecedented." To move "glacially" no longer implies slowness, and the "severe, widespread, and irreversible impacts" on people and nature that the most recent report of the Intergovernmental Panel on Climate Change (IPCC) warned us about have already come to fruition in Greenland.

It was in Qaanaaq in 1997 that I first experienced climate change from the feet up. I was traveling with Jens Danielsen, headed for Kiatak Island. It was spring, and six inches of snow covered the sea ice. Our 15 dogs trotted slowly; the only sound was their percussive panting. We had already encountered a series of pressure ridges —steep slabs of ice piled up between two floes—that took us five hours to cross. When we reached a smooth plain of ice again, we thought the worst was over, but the sound of something breaking shocked us: dogs began disappearing into the water. Jens hooked his feet over the front edge of the sled, lay on the trace lines, and pulled the dogs out. Afterward, he stepped down onto a piece of rotten ice, lifted the front of the sled, and laid it on a spot that was more stable, then jumped aboard and yelled at the dogs to run fast. When I asked if we were going to die, he smiled and said, "Imaqa." Maybe.

Ice-adapted people have amazing agility, which allows them to jump from one piece of drift ice to another and to handle half-wild dogs. They understand that life is transience, chance, and change. Because ice is so dynamic, melting in summer and re-forming in September, Greenlanders in the far north understand that nothing is solid, that boundaries are actually passages, that the world is a permeable place. On the ice they act quickly and precisely, flexing mind as well as muscle, always "modest in front of the weather," as Jens explained. Their material culture represents more than 10,000 years of use: dogsleds, kayaks, skin boats, polar-bear and sealskin pants, bone scrapers, harpoons, bearded seal–skin whips—all designed for beauty, efficiency, and survival in a harsh world where most people would be dead in a day.

From 1997 to 2012 I traveled by dogsled, usually with Jens and his three brothers-in-law: Mamarut Kristiansen, Mikile Kristiansen, and Gedeon Kristiansen. The dogtrot often lulled me to sleep, but rough ice shook me to attention. "You must look carefully," Jens said. From him I began to understand about being *silanigtalersarput:* a person who is wise about things and knows the ice, who comes to teach us how to see. The first word I learned in Greenlandic was *sila,* which means, simultaneously, weather, animal and human consciousness, and the power of nature. The Greenlanders I traveled with do not make the usual distinctions between a human mind and an animal mind. Polar bears are thought to understand human language. In the spring, mirages appear, lifting islands into the air and causing the ice to look like open water. Silver threads at the horizon mark the end of the known world and the beginning of the one inhabited by the imagination. Before television, the Internet, and cell phones arrived in Greenland, the coming of the dark time represented a shift: anxiety about the loss of light gave way to a deep, rich period of storytelling.

In Qaanaaq the sun goes down on October 24 and doesn't rise again until February 17. Once the hood of completely dark days arrives, with only the moon and snow to light the paths between houses, the old legends are told: "The Orphan Who Became a Giant," "The Orphan Who Drifted Out to Sea." Now Jens complains that the advent of television in Qaanaaq has reduced storytelling time, though only three channels are available. But out on the ice the old ways thrive. During the spring of 1998, when I traveled with Jens and his wife, Ilaitsuk, along with their five-year-old grandchild, installments of the legends were told to the child each night for two weeks.

That child, now a young man, did not become a subsistence hunter, despite his early training. He had seen too many springs when there was little ice. But no one suspected the ice would disappear completely.

The cycle of thinning and melting is now impossible to stop. The enormous ice sheet that covers 80 percent of the island is increasingly threaded with meltwater rivers in summer, though when I first arrived in Greenland, in 1993, it shone like a jewel. According to Konrad "Koni" Steffen, a climate scientist who has established many camps on top of the Greenland ice sheet, "in 2012 we lost 450 gigatons of ice—that's five times the amount of ice in the Alps. All the ice on top has pulled apart. It used to be smooth; now it looks like a huge hammer has hit it. The whole surface is fractured."

In 2004, with a generous grant from the National Geographic Expeditions Council, I returned to Qaanaaq for two monthlong journeys—in March and in July. The hunters had said to come in early March, one of the two coldest months in Greenland, because they were sure the ice would be strong then. They needed food for their families and their dogs. We would head south to Savissivik, a hard four-day trip. The last part would take us over the edge of the ice sheet and down a precipitous canyon to the frozen sea in an area they called Walrus El Dorado. It was –20 degrees when we started out with 58 dogs, four hunters—including Jens, Gedeon, Mamarut, and a relative of Jens's named Tobias—and my crew of three. We traveled on *hikuliaq*—ice that has just formed. How could it be only seven inches thick at this temperature? I asked Jens. He told me: "There is no old ice, it's all new ice and very salty: hard on the dogs' feet, and, you'll see, it melts fast. Dangerous to be going out on it." But there we were.

After making camp we walked single file to the ice edge. The ice was so thin that it rolled under our feet like rubber. One walrus was harpooned. It was cut up and laid on our sleds. I asked about the pile of intestines left behind. "That's for the foxes, ravens, and polar bears," Mamarut said. "We always leave food for others." Little did we know then that we would get only one walrus all month, and that soon we would be hungry and in need of meat for ourselves and the dogs.

The cold intensified and at the same time more ice broke up. We traveled all day in frigid temperatures that dropped to what Jens said was –40, and found refuge in a tiny hut. We spent the day rubbing ointment onto our frostbitten faces and fingers, and eating boiled walrus for hours at a time to keep warm. A day later we traveled south to Moriusaq, a village of 15, where the walrus hunting had always been good. But the ice there was unstable, too. We were told that farther south, around Savissivik, there was no ice at all. Mamarut's wife, Tekummeq, the great-granddaughter of the explorer Robert Peary, taught school in the village. She fed us and heated enough water for a bath. Finally we turned around and headed north toward Qaanaaq, four days away. Halfway there, a strong blizzard hit and we were forced to hole up in a hut for three days. We kept our visits outside brief, but after even a few minutes any exposed skin burned: fingers, hands, cheeks, noses, foreheads, and asses. The jokes flowed. The men kept busy fixing dog harnesses and sled runners. Evenings, they told hunting stories—not about who got the biggest animal but who made the most ridiculous mistake—to great laughter.

Days were white, nights were white. On the ice, dogs and humans eat the same food. The dogs lined up politely for the chunks of frozen walrus that their owners flung into their mouths. Inside the hut, a haunch of walrus hung from a hook, dripping blood. Our heat was a single Primus burner. Breakfast was walrus-heart soup; lunch was what Aleqa, our translator (who later became the first female prime minister of Greenland), called "swim fin"—a gelatinous walrus flipper. Jens, the natural leader of his family and the whole community, told of the polar bear with the human face, the one who could not be killed, who had asked him to follow, to become a shaman. "I said no. I couldn't desert my family and the community of hunters. This is the modern world, and there is no place in it for shamans."

When the temperature moderated, we spent three weeks trying to find ice that was strong enough to hold us. We were running out of food. The walrus meat was gone. Because Greenlandic freight sleds have no brakes, Jens used his legs and knees to slow us as we skidded down a rocky creekbed. At the bottom, we traveled down a narrow fjord. There was a hut and a drying rack: the last hunter to use the shed had left meat behind. The dogs would eat, but we would not—the meat was too old—and we were still a long way from home. The weather improved but it still averaged 30 degrees below zero. "Let's go out to Kiatak Island," Jens said. "Maybe we can get a walrus there." After crossing the strait, we traveled on an ice foot—a belt of ice that clung to the edge of the island. Where it broke off we had to unhook the dogs, push the sleds over a 14-foot cliff, and jump down onto rotting disks of ice. Sleds tipped and slid as dogs leaped over moats of open water from one spinning pane to the next. We traveled down the island's coast to another small hut, happy to have made it safely. From a steep mountain the men searched the frozen ocean for walrus with binoculars, but the few animals they saw were too far out and the path of ice to get to them was completely broken.

A boy from Siorapaluk showed up the next morning with a fine team, beautifully made clothing, a rifle, and a harpoon. At 15 he had taken a year off from school to see whether he had the prowess to be a great hunter, and he did. But the ice will not be there for him in the future; subsistence hunting will not be possible. "We weren't born to buy and sell things," Jens said sadly, "but to live with our families on the ice and hunt for our food."

Spring weather had come. The temperature had warmed considerably, and the air felt balmy. As we traveled to Siorapaluk, a mirage made Kiatak Island appear to float like an iceberg. Several times, while we stopped the dogs to rest, we stretched out on the sled in our polar-bear pants to bask in the warmth of the sun.

North of Siorapaluk there are no more habitations, but the men of the village go up the coast to hunt polar bears. When Gedeon and his older brother Mamarut ventured north for a few hours to see whether the route was an option for us, all they saw was a great latticed area of pressure ice, polynyas (perennially open water), and no polar bears. They decided against going farther. We had heavy loads, and the dogs had not eaten properly for a week, so after a rest at Siorapaluk we turned for home, traveling close to the coast on shore-fast ice.

On our arrival in Qaanaaq, the wives, children, and friends of the hunters greeted us and helped unload the sleds. The hunters explained that we had no meat. With up to 15 dogs per hunter, plus children, the sick, and the elderly, there were lots of mouths to feed. Northern Greenland is a food-sharing society with no private ownership of land. In these towns families own only the houses they build and live in, along with their dogs and their equipment. No one hunts alone; survival is a group effort. When things go wrong or the food supply dwindles, no one complains. They still have in their memories tales of hunger and famine. Greenland has its own government but gets subsidies from Denmark. In the old days, before the mid-1900s, an entire village could starve quickly, but now Qaanaaq has a grocery store, and with Danish welfare and help from extended families, no one goes without food.

Back in town after a month on the ice, we experienced "village shock." Instead of being disappointed about our failed walrus hunt, we celebrated with a bottle of wine and a wild dance at the local community hall, then talked until dawn. Finally my crew and I made our rounds of thanks and farewells and boarded the once-a-week plane south. It was the end of March, and just beginning to get warm. When I returned to Qaanaaq four months later, in July, the dogsleds had been put away, new kayaks were being built, and the edges of paddles were being sharpened to cut through roiling fjord water. I camped with the hunters' wives and children on steep hillsides and watched for pods of narwhals to swim up the fjord. "Qilaluaq," we'd yell when we saw a pod, enough time for Gedeon to paddle out and wait. As the narwhals swam by, he'd glide into the middle of them to throw a harpoon. By the end of the month enough meat had been procured for everyone. In August a hint of darkness began to creep in, an hour a day. Going back to Qaanaaq in Jens's skiff, I was astonished to see the moon for the first time in four months. Jens was eager to retrieve his dogs from the island where they ran loose all summer and to get out on the ice again, but because of the changing climate, the long months of darkness and twilight no longer marked the beginnings and endings of the traditional hunting season.

The year 2007 saw the warmest winter worldwide on record. I'd called the hunters in Qaanaaq that December to ask when I should come. It had been two years since I'd been there, and Jens was excited about going hunting together as we had when we first met. He said, "Come early in February when it's very cold, and maybe the ice will be strong." The day I arrived in Greenland I was shocked to find that it was warmer at the airport in Kangerlussuaq than in Boston. The ground crew was in shirtsleeves. I thought it was a joke. No such luck. Global air and sea temperatures were on the rise. The AO, the Arctic Oscillation, an index of high- and low-pressure zones, had recently switched out of its positive phase —when frigid air is confined to the Arctic in winter—and into its negative phase—when the Arctic stays warm and the cold air filters down into lower latitudes.

Flying north the next day to Qaanaaq, I looked down in disbelief: from Uummannaq, a village where I had spent my first years in Greenland, up to Savissivik, where we had tried to go walrus hunting, there was only open water threaded with long strings of rotting ice. As global temperatures increase, multiyear ice—ice that does not melt even in summer, once abundant in the High Arctic—is now disappearing. Finally, north of Thule Air Base and Cape York, ice had begun to form. To see white, and not the black ink of open water, was a relief. But that relief was short-lived. Greenland had entered what American glaciologist Jason Box calls "New Climate Land."

Jens, Mamarut, Mikile, and Gedeon came to the guesthouse when I arrived, but there was none of the usual merriment that precedes a long trip on the ice. Jens explained that only the shorefast ice was strong enough for a dogsled, that hunting had been impossible all winter. Despondent, he left. I heard rifle shots. What was that? I asked. "Some of the hunters are shooting their dogs because they have nothing to feed them," I was told. A 50-pound bag of dog food from Denmark cost more than the equivalent of 50 U.S. dollars; one bag lasts two days for 10 dogs.

Gedeon and Mikile offered to take me north to Siorapaluk. What was normally an easy 6-hour trip took 12 hours, with complicated pushes up and over an edge of the ice sheet. On the way, Gedeon recounted a narrow escape. He had gone out hunting against the better judgment of his older brother. His dogsled drifted out onto an ice floe that was rapidly disintegrating. He called for help. The message was sent to Thule Air Base, and a helicopter came quickly. Gedeon and the dogs (unhooked from the sled) were hauled up into the hovering aircraft. When he looked down, his dogsled and the ice on which he had been standing had disappeared.

We arrived at Siorapaluk late in the day, and the village was strangely quiet. It had once been a busy hub, with dogsleds coming and going, and polar-bear skins stretched out to dry in front of every house. There was a school, a chapel, a small store with a pay phone (from which you could call other Greenland towns), and a post office. Mail was picked up and delivered by helicopter; in earlier times, delivery of a letter sent by dogsled could take a year. Siorapaluk once was famous for its strong hunters who went north along the coast for walrus and polar bears. By 2007 everything had changed. There were almost no dog teams staked out on the ice, and quotas were being imposed on the harvest of polar bears and narwhals.

At the end of the first week I called a meeting of hunters so that I could ask them how climate change was affecting their lives. Otto Simigaq, one of the best Siorapaluk hunters, was eager to talk: "Seven years ago we could travel on safe ice all winter and hunt animals. We didn't worry about food then. Now it's different. There has been no ice for seven months. We always went to the ice edge in spring west of Kiatak Island, but the ice doesn't go out that far now. The walrus are still there, but we can't get to them." Pauline Simigaq, Otto's wife, said, "We are not so good in our outlook now. The ice is dangerous. I never used to worry, but now if Otto goes out I wonder if I will ever see him again. Around here it is depression and changing moods. We are becoming like the ice."

After the meeting I stood and looked out at the ruined ice. Beyond the village was Kiatak, and to the north was Neqe, where I had watched hunters climb straight up rock cliffs to scoop little auks, or dovekies, out of the air with long-handled nets. Farther north was the historic (now abandoned) site of Etah, the village where, in 1917, a half-starved Knud Rasmussen, returning from his difficult attempt to map the uninhabited parts of northern Greenland, came upon the American Crocker Land Expedition and the welcoming sound of a gramophone playing Wagner and Argentine tangos. Explorers and visitors came and went. Siorapaluk, Pitoravik, and Etah were regular stops for those going to the North Pole or to Ellesmere Island. Some, most notably Robert Peary, fathered children during their expeditions. The Greenlanders—and those children—stayed, traveling only as far as the ice took them. "We had everything here," Jens said. "Our entire culture was intact: our language and our way of living. We kept the old ways and took what we wanted of the new."

It wasn't until 2012 that I returned to Qaanaaq. I hadn't really wanted to go: I was afraid of what I would find. I'd heard that suicides and drinking had increased, that despair had become contagious. But a friend, the artist Mariele Neudecker, had asked me to accompany her to Qaanaaq so that she could photograph the ice. On a small plane carrying us north from Ilulissat she asked a question about glaciers, so I yelled out: "Any glaciologists aboard?" Three passengers, Poul Christoffersen, Steven Palmer, and Julian

Dowdeswell turned around and nodded. They hailed from Cambridge University's Scott Polar Research Institute and were on their way to examine the Greenland ice sheet north of Qaanaaq. As we looked down, Steve said, "With airborne radar we can identify the bed beneath several kilometers of ice." Poul added: "We're trying to determine the consequences of global warming for the ice." They talked about the linkages between ocean currents, atmosphere, and climate. Poul continued: "The feedbacks are complicated. Cold ice-sheet meltwater percolates down through the crevasses and flows into the fjords, where it mixes with warm ocean water. This mixing has a strong influence on the glaciers' flow."

Later in the year, they would present their new discovery: two subglacial lakes just north of Qaanaaq, half a mile beneath the ice surface. Although common in Antarctica, these deep hidden lakes had eluded glaciologists working in Greenland. Steve reported, "The lakes form an important part of the ice sheet's plumbing system connecting surface lakes to the ones beneath. Because the way water flows beneath ice sheets strongly affects ice-flow speeds, improved understanding of these lakes will allow us to predict more accurately how the ice sheet will respond to anticipated future warming."

Steve and Poul talked about four channels of warm seawater at the base of Petermann Glacier that allowed more ice islands to calve, and the 68-mile-wide calving front of the Humboldt Glacier, where Jens and I, plus seven other hunters, had tried to go one spring but were stopped when the dogs fell ill with distemper and died. Even with healthy dogs we wouldn't be able to go there now. Poul said that the sea ice was broken and dark jets of water were pulsing out from in front of the glacier—a sign that surface and subglacial meltwater was coming from the base of the glacier, exacerbating the melting of the ice fronts and the erosion of the glacier's face.

The flight from Ilulissat to Qaanaaq takes three hours. Below us, a cracked elbow of ice bent and dropped, and long stretches of open water made sparkling slits cuffed by rising mist. Even from the plane we could see how the climate feedback loop works, how patches of open water gather heat and produce a warm cloud that hangs in place so that no ice can form under it. "Is it too late to rewrite our destiny, to reverse our devolution?" I asked the glaciologists. No one answered. We stared at the rotting ice. It was down there that a modern shaman named Panippaq, who was said to be capable of heaping up mounds of fish at will, had committed suicide as he watched the sea ice decline. Steve reminded me that the global concentration of carbon dioxide in the atmosphere had almost reached 400 parts per million, and that the Arctic had warmed at least five degrees. Julian Dowdeswell, the head of the institute at Cambridge, had let the younger glaciologists do the talking. He said only this: "It's too late to change anything. All we can do now is deal with the consequences. Global sea level is rising."

But when Mariele and I arrived in Qaanaaq, we were pleasantly surprised to find that the sea ice was three feet thick. Narwhals, beluga, and walrus swam in the leads of open water at the ice edge. Pairs of eider ducks flew overhead, and little auks arrived by the thousands to nest and fledge in the rock cliffs at Neqe. Spirits rose. I asked Jens whether they'd ever thought of starting a new community farther north. He said they had tried, but as the ice retreated hungry polar bears had come onto the land, as they were doing in Vankarem, Russia, and Kaktovik, Alaska. The bears were very aggressive. "We must live as we always have with what the day brings to us. And today, there is ice," he said.

Jens had recently been elected mayor of Qaanaaq and had to leave for a conference in Belgium, but Mamarut, Mikile, and Gedeon wanted to hunt. When we went down to the ice where the dogs were staked, I was surprised to see Mikile drunk. Usually mild-mannered and quiet, he lost control of his dogs before he could get them hitched up, and they ran off. With help from another hunter, it took several hours to retrieve them. Perched on Mikile's extra-long sled was a skiff; Mamarut tipped his kayak sideways and lashed it to his sled. Gedeon carried his kayak, paddles, guns, tents, and food on his sled, plus his new girlfriend, Bertha. The spring snow was wet and the going was slow, but it was wonderful to be on a dogsled again.

I had dozed off when Mamarut whispered, "Hiku hina," in my ear. The ice edge. Camp was set up. Gedeon sharpened his harpoon, and Bertha melted chunks of ice over a Primus stove for tea. The men carried their kayaks to the water's edge. Glaucous gulls flew by. The sound of narwhal breathing grew louder. "Qilaluaq!" Gedeon whispered. The pod swam by but no one went after them. It was May, and the sun was circling in a halo above our heads, so we learned to sleep in bright light. It was time to rest. We laid our sleeping bags under a canvas tent, on beds made from two sleds pushed together. The midnight sun tinted the sea green, pink, gray, and pale blue.

Hours later, I saw Gedeon and Mikile kneeling in snow at the edge of the ice, facing the water. They were careful not to make eye contact with passing narwhals: two more pods had come by, but the men didn't go after them. "They have too many young ones," Gedeon whispered, before continuing his vigil. Another pod approached and Gedeon climbed into his boat, lithe as a cat. He waited, head down, with a hand steadying the kayak on the ice edge. There was a sound of splashing and breathing, and Gedeon exploded into action, paddling hard into the middle of the pod, his kayak thrown around by turbulent water. He grabbed his harpoon from the deck of the kayak and hurled it. Missed. He turned, smiling, and paddled back to camp. There was ice and there was time—at least for now—and he would try again later.

In the night, a group of Qaanaaq hunters arrived and made camp behind us on the ice. It's thought to be bad practice to usurp another family's hunting area. They should have moved on but didn't. No one said anything. The old courtesies were disintegrating along with the ice. The next morning, a dogfight broke out, and an old man viciously beat one of his dogs with a snow shovel. In 20 years of traveling in Greenland, I'd never seen anyone beat a dog.

Hunting was good the next day, and the brothers were happy to have food to bring home for their families. Though the ice was strong, they knew better than to count on anything. We were all deeply upset about the beating we had witnessed, but there was nothing we could do. In Greenland there are unwritten codes of honor that, together with the old taboos, have kept the society humming. A hunter who goes out only for himself and not for the group will be shunned: if he has trouble on the ice, no one will stop to help him. Hunters don't abuse their dogs, which they rely on for their lives.

To become a subsistence hunter, the most honorable occupation in this society, is no longer an option for young people. "We may be coming to a time when it is summer all year," Mamarut said as he mended a dog harness. Once the strongest hunter of the family and also the jokester, he was now too banged up to hunt and rarely smiled. He'd broken his ankle going solo across the ice sheet in a desperate attempt to find food—hunting muskoxen instead of walrus—and it took him two weeks to get home to see a doctor. Another week went by before he could fly to Nuuk, the capital of Greenland, for surgery. Now the ankle gives him trouble and his shoulder hurts: one of his rotator cuffs is torn. The previous winter his mother died—she was still making polar-bear pants for her sons, now middle-aged—and a fourth brother committed suicide. "They want us to become fishermen," Mamarut said. "How can we be something we are not?"

On the last day we camped at the ice edge, the hunters got 2 walrus, 4 narwhals, and 10 halibut. As the men paddled back to camp, their dogs broke into spontaneous howls of excitement. Mamarut had opted to stay in camp and begin packing. In matters of hunting, his brash younger brother, Gedeon, had taken his place. Eight years earlier I had watched Gedeon teach his son, Rasmus, how to handle dogs, paddle a kayak, and throw a harpoon. Rasmus was seven at the time. Now he goes to school in south Greenland, below the Arctic Circle, and is learning to be an electrician. Mamarut and his wife, Tekummeq, have adopted Jens and Ilaitsuk's grandchild, but rather than being raised in a community of traditional hunters, the child will grow up on an island nation whose perennially open waters will prove attractive to foreign oil companies.

At camp, Mamarut helped his two brothers haul the dead animals onto the ice. One walrus had waged an urgent fight after being harpooned and had attacked the boat. Unhappy that the animal did not die instantly, Gedeon had pulled out his rifle and fired, ending the struggle that was painful to watch. The meat was butchered in silence and laid under blue tarps on the dogsleds. Breakfast was fresh narwhal-heart soup, rolls with imported Danish honey, and *mattak*—whale skin, which is rich in vitamin C, essential food in an environment that can grow no fruits or vegetables.

We packed up camp, eager to leave the dog beater behind. It was the third week of May and the temperature was rising: the ice was beginning to get soft. We departed early so that the three-foot gap in the ice that we had to cross would still be frozen, but as soon as the sun appeared from behind the clouds, it turned so warm that we shed our anoraks and sealskin mittens. "Tonight that whole ice edge where we were camped will break off," Mamarut said quietly. The tracks of *ukaleq* (Arctic hare) zigzagged ahead of us, and Mamarut signaled to the dogs to stay close to the coast lest the ice on which we were traveling break away. We camped high on a hill in a small hut near the calving face of Politiken's Glacier, which in 1997 had provided an easy route to the ice sheet but was now a chaos of rubble. Mamarut laid out the topographic map I had brought to Greenland on my first visit, in 1993, and scrutinized the marks we had made over the years showing the ice's retreat. Once the ice edge in the spring extended far out into the strait; now it barely reached beyond the shore-fast ice of Qaanaaq. Despite seasonal fluxes, the ice kept thinning. Looking at the map, Mamarut shook his head in dismay. "Ice no good!" he blurted out in English, as if it were the best language for expressing anger. On our way home to Qaanaaq the next day, he got tangled in the trace lines while hooking up the dogs and was dragged for a long way before I could stop them. These were the final days of subsistence hunting on the ice, and I wondered if I would travel with these men ever again.

The news from the Ice Desk is this: the prognosis for the future of Arctic ice, and thus for human life on the planet, is grim. In the summer of 2013 I returned to Greenland, not to Qaanaaq but to the town of Ilulissat in what's known as West Greenland, the site of the Jakobshavn Glacier, the fastest-calving glacier in the world. I was traveling with my husband, Neal, who was on assignment to produce a radio segment on the accelerated melting of the Greenland ice sheet. In Copenhagen, on our way to Ilulissat, we met with Jason Box, who had moved to Denmark from the prestigious Byrd Polar and Climate Research Center to work in Greenland. It was a sunny Friday afternoon, and we agreed to meet at a canal where young Danes, just getting off work, piled onto their small boats, to relax with a bottle of wine or a few beers. Jason strolled toward us wearing shorts and clogs, carrying a bottle of hard apple cider and three glasses. His casual demeanor belies a gravity and intelligence that becomes evident when he talks. A self-proclaimed climate refugee, and the father of a young child, he said he couldn't live with himself if he didn't do everything possible to transmit his understanding of abrupt climate change in the Arctic and its dire consequences.

Jason has spent 24 summers atop Greenland's great dome of ice. "The ice sheet is melting at an accelerated pace," he told us. "It's not just surface melt but the deformation of the inner ice. The fabric of the ice sheet is coming apart because of increasing meltwater infiltration. Two to three hundred billion tons of ice are being lost each year. The last time atmospheric CO_2 was this high, the sea level was seventy feet higher."

We flew to Ilulissat the next day. Below the plane, milky-green water squeezed from between the toes of glaciers that had oozed down from the ice sheet. Just before landing, we glided over a crumpled ribbon of ice that was studded with icebergs the size of warehouses: the fjord leading seaward from the calving front of the Jakobshavn Glacier. Ice there is moving away from the central ice sheet so fast—up to 150 feet a day—and calves so often that the adjacent fjord has been designated a World Heritage Site, an ironic celebration of its continuing demise. Ilulissat was booming with tourists who had flocked to town to observe the parade of icebergs drift by as they sipped cocktails and feasted on barbecued muskoxen at the four-star Hotel Arctic; it was also brimming with petroleum engineers who had come in a gold-rush-like flurry to find oil. But the weather had changed: many of the well sites were nonproducers, and just below the fancy hotel were the remains of several tumbled houses and a ravine that had been dredged by a flash flood, a rare weather event in a polar desert.

Neal and I hiked up the moraine above town to look down on the ice-choked fjord. We sat on a promontory to watch and listen to the ice pushing into Disko Bay. Nothing seemed to be moving, but at the front of stranded icebergs fast-flowing streams of meltwater spewed out, crisscrossing one another in the channel. Recently several subglacial lakes were discovered to have "blown out," draining as much as 57,000 gallons per minute and then refilling with surface meltwater, softening the ice around it, so that the entire ice sheet is in a process of decay. From atop another granite cliff we saw an enormous berg, its base smooth but its top all jagged with pointed slabs. Suddenly, two thumping roars, another sharp thud, and an entire white wall slid straight down into the water. Neal turned to me, wide-eyed, and said: "This is the sound of the ice sheet melting."

Later, we gathered at the Hotel Icefiord with Koni Steffen and a group of Dartmouth glaciology students. Under a warm sun we sat on a large deck and discussed the changes that have occurred in the Arctic in the past five years. Vast methane plumes were discovered boiling up from the Laptev Sea, north of Russia, and methane is punching through thawing seabeds and terrestrial permafrost all across the Arctic. Currents and air temperatures are changing; the jet stream is becoming wavier, allowing weather conditions to persist for long periods of time; and the movements of high- and low-pressure systems have become unpredictable. The new chemical interplay between ocean and atmosphere is now so complex that even Steffen, the elder statesman of glaciology, says that no one fully understands it. We talked about future scenarios of what we began to call, simply, bad weather. Parts of the world will get much hotter, with no rain or snow at all. In western North America, trees will keep dying from insect and fungal invasions, uncovering more land that in turn will soak up more heat. It's predicted that worldwide demand for water will exceed the supply by 40 percent. Cary Fowler, who helped found the Svalbard Global Seed Vault, predicts that there will be such dire changes in seasonality that food growing will no longer align with rainfall, and that we are not prepared for worsening droughts. Steffen says, "Water vapor is now the most plentiful and prolific greenhouse gas. It is altering the jet stream. That's the truth, and it shocks all the environmentalists!"

In a conversation with the biologist E. O. Wilson on a morning in Aspen so beautiful that it was difficult to imagine that anything on the planet could go wrong, he advised me to stop being gloomy. "It's our chance to practice altruism," he said. I looked at him skeptically. He continued: "We have to wear suits of armor like World War II soldiers and just keep going. We have to get used to the changes in the landscape, to step over the dead bodies, so to speak, and discipline our behavior instead of getting stuck in tribal and religious restrictions. We have to work altruistically and cooperatively, and make a new world."

Is it possible we haven't fully comprehended that we are in danger? We may die off as a species from mere carelessness. That night in Ilulissat, on the patio of the Hotel Icefiord, I asked one of the graduate students about her future. She said: "I won't have children; I will move north." We were still sitting outside when the night air turned so cold that we had to bundle up in parkas and mittens to continue talking. "A small change can have a great effect," Steffen said. He was referring to how carelessly we underestimate the profound sensitivity of the planet's membrane, its skin of ice. The Arctic has been warming more than twice as fast as anywhere else in the world, and that evening, the reality of what was happening to his beloved Greenland seemed to make Steffen go quiet. On July 30, 2013, the highest temperature ever recorded in Greenland—almost 80 degrees Fahrenheit—occurred in Maniitsoq, on the west coast, and an astonishing heat wave in the Russian Arctic registered 90 degrees. And that was 2013, when there was said to be a "pause" in global heating.

Recently, methane plumes were discovered at 570 places along the East Coast of the United States, from Cape Hatteras, North Carolina, to Massachusetts. Siberian tundra holes were spotted by nomadic reindeer herders on the Yamal Peninsula, and ash from wildfires in the American and Canadian West fluttered down, turning the southern end of the Greenland ice sheet almost black.

The summer after Neal and I met with Koni Steffen in Ilulissat, Jason Box moved his camp farther north, where he continued his attempts to unveil the subtle interactions between atmosphere and earth, water, and ice, and the ways algae and industrial and wildfire soot affect the reflectivity of the Greenland ice sheet: the darker the ice, the more heat it absorbs. As part of his recent Dark Snow Project, he used small drones to fly over the darkening snow and ice. By the end of August 2014, Jason's reports had grown increasingly urgent. "We are on a trajectory to awaken a runaway climate heating that will ravage global agricultural systems, leading to mass famine and conflict," he wrote. "Sea-level rise will be a small problem by comparison. We simply must lower atmospheric carbon emissions." A later message was frantic: "If even a small fraction of Arctic seafloor methane is released to the atmosphere, we're fucked." From an IPCC meeting in Copenhagen last year, he wrote: "We have very limited time to avert climate impacts that will ravage us irreversibly."

The Arctic is shouldering the wounds of the world, wounds that aren't healing. Long ago we exceeded the carrying capacity of the planet, with its seven billion humans all longing for some semblance of First World comforts. The burgeoning population is incompatible with the natural economy of biological and ecological systems. We have found that our climate models have been too conservative, that the published results of science-by-committee are unable to keep up with the startling responsiveness of Earth to our every footstep. We have to stop pretending that there is a way back to the lush, comfortable, interglacial paradise we left behind so hurriedly in the 20th century. There are no rules for living on this planet, only consequences. What is needed is an open exchange in which sentience shapes the eye and mind and results in ever-deepening empathy. Beauty and blood and what Ralph Waldo Emerson called "strange sympathies" with otherness would circulate freely in us, and the songs of the bearded seal's ululating mating call, the crack and groan of ancient ice, the Arctic tern's cry, and the robin's evensong would inhabit our vocal cords.

ROSE EVELETH

Why Are Sports Bras So Terrible?

FROM *Racked*

THE FIRST THING TO KNOW about sports and breasts is this: women have always participated in athletics, bra or no bra. In ancient Rome, women bound their breasts with cloth and leather. Pottery and mosaics from the fourth and fifth centuries show female athletes wearing bikini-like uniforms.

In the Victorian era, women turned to corsets to keep their breasts from moving too much. Those competing at Wimbledon in 1887 returned to their dressing rooms in between matches to "unhitch their bloody corsets," having been "repeatedly stabbed by the metal and whale bone stays of the cumbersome garments" as they played.

By 1911 women got a "sports corset" with flexible material, and thanks to the 1914 tango craze, someone even invented a dancing corset. But it wasn't until the 1920s that bras started to replace corsets in the United States, and while brassieres designed for athletic purposes were patented as early as 1906, they simply never caught on.

Finally, in 1977—the same year Victoria's Secret was founded—the sports bra as we know it was invented by Lisa Lindahl and Polly Smith, with the help of designer and runner Hinda Miller. That first sports bra was simply two jockstraps sewn together. It wasn't just that jockstraps were the right size, they were also the right idea. "We said, what we really need to do is what men have been doing: pull everything close to the body," Miller later told researchers. They called this new bra the Jockbra, but quickly changed it to Jogbra after store owners in South Carolina deemed the name offensive.

During its first year on the market, Jogbra moved 25,000 units. Two decades later, in 1998, the sports bra industry sold $412 million worth of product. A 2002 study estimated that sports bras accounted for about 6 percent of the then-$4.5 billion bra market. Today, the bra market is worth about $15 billion. Factor in that female participation in sports is increasing every year and that "athleisure" appears to be here to stay, and it's no wonder that from Lululemon to Under Armour to Victoria's Secret, brands are turning their attention to sports bras.

But researchers are still a long way from understanding exactly how breasts move during exercise. Standing in the way of designing the best sports bra possible is millennia of stigma, powerful marketing forces, and good old-fashioned physics.

Breasts have no muscle. They sit on top of the pectoral muscles, but breasts themselves are all fat and glands and connective tissue. They're held to the chest by something called Cooper's ligaments, though those ligaments aren't designed to reduce movement. As one study puts it, "the skin appears to provide most of the support for the breast in regards to limiting breast movement."

That is to say that there is nothing biological working to stop breasts from moving. Without any such built-in support, as any person with breasts can attest, they bounce up and down freely, which can cause a fair bit of discomfort. When surveyed, between 40 and 60 percent of women report breast pain associated with physical activity. That pain makes women less likely to exercise and, among those who do, hurts their performance.

Understanding the biomechanics of bouncing is key to understanding how to make it stop, but it's a field that's only recently gained traction. And since breast size, placement, and density are different for every woman, researchers need to look at a large sample to get a good idea of what's going on.

Even two people with the same size breasts might have different breast composition; put them in the same bra, and one might be in heaven while the other can barely breathe. "There are so many factors going on, it's hard to pin down that it's the bra," says Jenny White, a researcher at the University of Portsmouth who studies breast motion.

White's research aims to better understand how breasts move, and what that means for the people who have them. When volunteers come into her lab, she has them do a variety of physical activities in a variety of bras (and without one), and asks them to report how each activity and fit feels. She also uses a sophisticated motion-capture system, placing reflective markers on the bra.

Recruiting for this kind of study can be hard, White says, but she came up with a clever strategy. In 2013 her team targeted female runners in the London Marathon: "At registration, we tried to accost as many of them as possible. We got about thirteen hundred people and we were able to understand that population."

It's one thing to be able to see a woman run in a bra in a lab. It's another to see what they experience after 26.2 miles. Over the course of a marathon, White says, "you might start seeing changes in the patterns of their running. You're probably going to see a decrease in stride levels. You're just not performing as well as you could."

On the other side of the world, at the University of Wollongong in Australia, Julie Steele does similar research. Women who sign up for Steele's studies come in and are all given the same standard-issue, commercially available bra. This lets the team compare how women fit themselves, how they're fitted by experts, and how they fill out and move in the control bra.

They then go through a series of tests to figure out the mechanics. "We use 3-D scanning, ultrasounds, and devices that measure the skin to really better understand the structure of the breasts of these women," says Steele. "We really need to better understand how that goes together to make the breasts move."

Both Steele and White say there's a constant battle going on between controlling the bounce and making bras comfortable. You could squash the breasts tightly against the body, but that makes it hard to breathe. The most common complaint, in the long list of ways in which sports bras are painful, is that the straps are too tight, sometimes even causing them to break.

It's worth noting that wearing the wrong sports bra isn't just a matter of discomfort or annoyance. Studies have shown that breast discomfort is a leading reason women stop participating in sports. And in extreme cases, an ill-fitting bra can actually do nerve damage. Bra straps generally cross over the brachial plexus, the nerve bundle that sends impulses to and from the arm. Women who wear bras with too-tight straps can damage that bundle, causing pain and numbness.

Another thing Steele and White worry about is that women aren't wearing the right size bra. A full 75 percent of the marathon runners White talked to had some kind of bra problem during training. "I think we need to scrap the whole bra system and start again from a sizing system," Steele says. Bra sizing is confusing, imprecise, and variable. This isn't just a sports bra problem, either; some surveys say that literally 100 percent of women are wearing the wrong size bra.

White says that the methods she, Steele, and their peers use to study breast movement might soon change dramatically. "We're at this crossover period," she says. White thinks that in the near future, they'll be using something called inertial sensors—tiny sensors that can be placed directly on the breast itself to gather GPS and acceleration data.

There are a few benefits to this new method. First, the sensors that attach to the breast directly rather than to the sports bra can tell researchers much more information about what the breast is doing underneath the bra. "So it means we completely get rid of any discrepancy between what the breast is doing and what the bra is doing," she says. In other words, if the bra is too big, or has padding or anything else that might make it move differently from how the breast itself is moving, the researchers will be able to detect that.

Second, these sensors don't require the camera setup that White's markers do, which means researchers could use them out in the field and see how breasts move in their natural environments, like during a soccer game or in a road race.

The sensor technology isn't quite there yet, but White says that it's coming: "It's been a slow development, they haven't been small enough to place on the breast. But they're getting that small, so we've got a provisional system."

Today, there are a lot more choices than the original Jogbra jockstrap design. In fact, as anybody who has recently gone shopping for a sports bra can attest, there is an overwhelming number of choices, from strappy yoga designs to padded cups to the classic racerback. But the choices women face come down to two main categories: compression bras and encapsulation bras.

Compression bras are the bras most people associate with sports bras—a single panel of fabric that hugs the breasts into the chest. The idea here is that if you can compress the breasts against the body, pulling them so they're closer to your center of gravity, they'll bounce less. Which is true, as long as the breasts in question aren't too large.

Encapsulation bras treat each breast individually, more like a regular bra. While compression bras work perfectly well for women who fall into the A- and B-cup range, larger-breasted women need more support. Some studies suggest that encapsulation bras can provide that, but not everyone is in agreement. In 2009 White found that for D-cup women, the difference in breast displacement between a compression bra and an encapsulation bra was insignificant.

"Once you go above a certain size, you start coming out everywhere anyway," White explains. She also notes that compression bras are far more common in the United States than elsewhere in the world. For very large-breasted women, some bras combine the two strategies, using individual cups for each breast with an outer compression layer on top.

In October of last year, Nike announced a new line of athletic gear for women, including sports bras, that it hoped would bring in $2 billion annually by 2017. This year, Under Armour announced its plans to reach $7.5 billion in sales, and as part of that growth, executives have pointed to its growing sports bra line.

It's not just athletic brands getting in on the action, either; designers of all stripes are dipping their toes into the athletic market. Designer Mara Hoffman just released an activewear line that includes colorful sports bras. Rebecca Minkoff and Tory Burch recently did the same. There's even a startup trying to disrupt the sports bra industry.

While there are suddenly a ton of options out there for women, not all of them are good. "There is no piece of clothing that is more difficult to design well than a sports bra," says LaJean Lawson, a breast researcher and consultant for Champion. "There are so many different parameters. It's the most hooked-into cultural stereotypes. You have to think about sweat, support, chafing, straps, slippage, and then looking cute. That's a really long list of conflicting design requirements."

Lawson has been studying breasts, their movement, and the

bras that contain them for Champion since 1984, when she tested the original Jogbra. In 1987 she evaluated the seven sports bras Champion had on offer using a 16 mm film camera. Today, she employs the same methods that White's lab does, using complex camera tracking to test hundreds of bras both from Champion and other companies. "I just ordered a thousand dollars' worth of Victoria's Secret sports bras to test in the lab," Lawson says, "and they called me and asked, 'Who are you, and why are you buying all these bras?'"

White's team works directly with brands that want to evaluate their bras using her methods. She runs breast-science workshops a few times a year, where representatives get an introductory crash course and learn ways bras can be designed to help reduce pain. From there, White offers her lab's services to companies that want to come in and test their designs.

One of the brands that has taken White up on the offer is Shock Absorber, a UK sports bra company that prides itself on its scientifically supported designs. "Since this original research was carried out, Shock Absorber has tested all new styles at the University of Portsmouth to measure their reduction in breast movement," its site states.

White wouldn't share the names of other brands she's worked with, but Kelly Cortina, vice president of women's apparel at Under Armour, disclosed that her team has used the Portsmouth lab. There, Cortina says, "we tested on women of all different sizes to ensure that the bra is minimizing breast movement and managing moisture efficiently." White's lab offers everything from a basic bra test to a "gold-level" setup where the company gets heavily involved in the fundamental research the team is doing.

In her ideal world, White would work with brands from the very beginning of their design phase, so they can test out many iterations of their styles. But that's not how it usually happens: "Unfortunately many companies have gone through the whole process of design and they just want to know how good it is."

Lawson, on the other hand, gets to work on Champion's designs from the start. Along with a team of engineers, Lawson advises on seam placement, strap design, how to change contours, which parts will rub, and more.

White and Lawson keep a close eye on the trends in sports bras, and they have both seen a recent rise in bras with padding. "Last spring, when I did my last project, I had twelve bras to test," Lawson says. "Four were Champion, eight were competitors, and ten out of those twelve had some kind of padding in them." White also saw more bras featuring lots of straps, and ones that look more like everyday bras: "I've noticed more sports bras that we're testing having underwire."

Here is where those athleisure offerings from brands like Mara Hoffman and Tory Burch are having an impact. "It's had a noticeable influence on what women want in their sports bra," Lawson says. "There have never been more products out on the market from nonathletic brands that don't have the structure and the design to meet women's needs when she's out kicking it hard on the road." Lawson adds that when she tests those bras on women in the lab, their flaws quickly become apparent. "When I can't get a woman to take an eighty-dollar sports bra home for free, I mean, that's a bad sign."

But this is the really tricky part: how good a sports bra is depends on what you're measuring. For White, the quality of a bra has everything to do with the comfort of the user wearing it. But for brands, that's not the only consideration.

Sports bras aren't just a piece of sports equipment. They're not like a bat or a baseball mitt or shin guards—designed, for the most part, for maximum functionality. They're cultural objects, they're fashion objects, and as such they're laden with all kinds of baggage about how a woman is supposed to look. "There's so much more to a sports bra than just a bra," says Jaime Schultz, a sociologist who studies women and sports.

Just look at the way sports bras are advertised. The very first ad for the Jogbra boasted that its "unique design holds breasts close to the body." Twenty years later, the company changed its tune. Suddenly, holding breasts close to the body, literally the entire purpose of the Jogbra, was a no-no. "Only abs should be flat," a new ad read in 1996. "Now, a sports bra that respects and defines your natural shape."

Today, Victoria's Secret is continuing the war against compressed breasts. They call this "the uniboob." Victoria's Secret chief executive officer Sharen Jester Turney announced last year that the war against the uniboob was in full force: "We wanted to solve the uniboob problem, where your sports bra makes you look straight across—no one likes that. This bra is just as much about performance and function as the look."

Well guess what, sometimes the uniboob is in fact the best way to reduce breast motion and pain. But women are constantly being told that even their sports bra should be sexy. Styles with spaghetti straps and low Vs and padded cups win out over wide straps and good support. "Women feel like they have to present themselves in the best possible breasted way that will appeal sexually," says Schultz.

Under Armour's Cortina echoed some of those sentiments in an email. "Women want to feel good and look good," she wrote. "Gone are the days of sacrificing style for fit or comfort. She expects and DESERVES a bra (no matter the cup size) to fit, perform, and be on trend. Many bras offer one but not the others. She wants it all and she can have it all!"

Lawson says that she hears from lots of women that they want padding, and while she herself is not a proponent of enhanced sports bras, she's also quick to say she doesn't want to dismiss other women's desires: "One of my testers is an aspiring Olympic marathoner, she's a 32C and she's like, 'I don't want to be flat.' That shocked me, but that's one of those things that has turned my ideas on their ear in the last few years."

This is a perfect example of the "be everything at once" dilemma women face. A bra can't just be a good bra, it also has to be fashionable and womanly. It has to hold the girls nicely, without diminishing their size and shape. And when forced to choose between those two things, bra manufacturers almost invariably choose look over function.

When asked about the future of sports bras, Lawson talked more about materials than overall design. She expects we'll see materials "that can respond to breast impact so the control is local rather than being on the shoulder, materials that are better at responding to body temperatures and heat rates." The compression and encapsulation styles, she says, are serviceable enough. While they could certainly use improvements in form, the foundational system of sports bras probably won't change much anytime soon.

And therein lies the problem. Lawson or White or any bra designer could, tomorrow, invent the world's best sports bra. Something comfortable and supportive and soft and easy to put on and take off. Something that wicks away sweat while providing coverage. Something that doesn't pinch the shoulders or squeeze the rib cage. But if that sports bra isn't cute, it wouldn't matter. "I think the biggest problem is that a lot of it isn't based on science," says Steele. "It's based on fashion and look. And that's what sells."

AMANDA GEFTER

The Man Who Tried to Redeem the World with Logic

FROM *Nautilus*

WALTER PITTS WAS USED to being bullied. He'd been born into a tough family in Prohibition-era Detroit, where his father, a boilermaker, had no trouble raising his fists to get his way. The neighborhood boys weren't much better. One afternoon in 1935, they chased him through the streets until he ducked into the local library to hide. The library was familiar ground, where he had taught himself Greek, Latin, logic, and mathematics—better than home, where his father insisted he drop out of school and go to work. Outside, the world was messy. Inside, it all made sense.

Not wanting to risk another run-in that night, Pitts stayed hidden until the library closed for the evening. Alone, he wandered through the stacks of books until he came across *Principia Mathematica,* a three-volume tome written by Bertrand Russell and Alfred Whitehead between 1910 and 1913, which attempted to reduce all of mathematics to pure logic. Pitts sat down and began to read. For three days he remained in the library until he had read each volume cover to cover—nearly 2,000 pages in all—and had identified several mistakes. Deciding that Bertrand Russell himself needed to know about these, the boy drafted a letter to Russell detailing the errors. Not only did Russell write back, he was so impressed that he invited Pitts to study with him as a graduate student at Cambridge University in England. Pitts couldn't oblige him, though—he was only 12 years old. But three years later, when he heard that Russell would be visiting the University of Chicago, the 15-year-old ran away from home and headed for Illinois. He never saw his family again.

In 1923, the year that Walter Pitts was born, a 25-year-old Warren McCulloch was also digesting the *Principia.* But that is where the similarities ended—McCulloch could not have come from a more different world. Born into a well-to-do East Coast family of lawyers, doctors, theologians, and engineers, McCulloch attended a private boys' academy in New Jersey, then studied mathematics at Haverford College in Pennsylvania, then philosophy and psychology at Yale. In 1923 he was at Columbia, where he was studying "experimental aesthetics" and was about to earn his medical degree in neurophysiology. But McCulloch was a philosopher at heart. He wanted to know what it means to know. Freud had just published *The Ego and the Id,* and psychoanalysis was all the rage. McCulloch didn't buy it—he felt certain that somehow the mysterious workings and failings of the mind were rooted in the purely mechanical firings of neurons in the brain.

Though they started at opposite ends of the socioeconomic spectrum, McCulloch and Pitts were destined to live, work, and die together. Along the way, they would create the first mechanistic theory of the mind, the first computational approach to neuroscience, the logical design of modern computers, and the pillars of artificial intelligence. But this is more than a story about a fruitful research collaboration. It is also about the bonds of friendship, the fragility of the mind, and the limits of logic's ability to redeem a messy and imperfect world.

Standing face to face, they were an unlikely pair. McCulloch, 42 years old when he met Pitts, was a confident, gray-eyed, wild-bearded, chain-smoking philosopher-poet who lived on whiskey and ice cream and never went to bed before 4:00 a.m. Pitts, 18, was small and shy, with a long forehead that prematurely aged him, and a squat, ducklike, bespectacled face. McCulloch was a respected scientist. Pitts was a homeless runaway. He'd been hanging around the University of Chicago, working a menial job and sneaking into Russell's lectures, where he met a young medical student named Jerome Lettvin. It was Lettvin who introduced the two men. The moment they spoke, they realized they shared a hero in common: Gottfried Leibniz. The 17th-century philosopher had attempted to create an alphabet of human thought, each letter of which represented a concept and could be combined and manipulated according to a set of logical rules to compute all knowledge —a vision that promised to transform the imperfect outside world into the rational sanctuary of a library.

McCulloch explained to Pitts that he was trying to model the brain with a Leibnizian logical calculus. He had been inspired by the *Principia,* in which Russell and Whitehead tried to show that all of mathematics could be built from the ground up using basic, indisputable logic. Their building block was the proposition —the simplest possible statement, either true or false. From there, they employed the fundamental operations of logic, like the conjunction ("and"), disjunction ("or"), and negation ("not"), to link propositions into increasingly complicated networks. From these simple propositions, they derived the full complexity of modern mathematics.

Which got McCulloch thinking about neurons. He knew that each of the brain's nerve cells only fires after a minimum threshold has been reached: enough of its neighboring nerve cells must send signals across the neuron's synapses before it will fire off its own electrical spike. It occurred to McCulloch that this setup was binary—either the neuron fires or it doesn't. A neuron's signal, he realized, is a proposition, and neurons seemed to work like logic gates, taking in multiple inputs and producing a single output. By varying a neuron's firing threshold, it could be made to perform "and," "or," and "not" functions.

Fresh from reading a new paper by a British mathematician named Alan Turing which proved the possibility of a machine that could compute any function (so long as it was possible to do so in a finite number of steps), McCulloch became convinced that the brain was just such a machine—one which uses logic encoded in neural networks to *compute.* Neurons, he thought, could be linked together by the rules of logic to build more complex chains of thought, in the same way that the *Principia* linked chains of propositions to build complex mathematics.

As McCulloch explained his project, Pitts understood it immediately, and knew exactly which mathematical tools could be used. McCulloch, enchanted, invited the teen to live with him and his family in Hinsdale, a rural suburb on the outskirts of Chicago. The Hinsdale household was a bustling, free-spirited bohemia. Chicago intellectuals and literary types constantly dropped by the house to discuss poetry, psychology, and radical politics while Spanish Civil War and union songs blared from the phonograph. But late at night, when McCulloch's wife, Rook, and the three children went to bed, McCulloch and Pitts alone would pour the whiskey, hunker down, and attempt to build a computational brain from the neuron up.

Before Pitts's arrival, McCulloch had hit a wall: there was nothing stopping chains of neurons from twisting themselves into loops, so that the output of the last neuron in a chain became the input of the first—a neural network chasing its tail. McCulloch had no idea how to model that mathematically. From the point of view of logic, a loop smells a lot like paradox: the consequent becomes the antecedent, the effect becomes the cause. McCulloch had been labeling each link in the chain with a time stamp, so that if the first neuron fired at time t, the next one fired at $t + 1$, and so on. But when the chains circled back, $t + 1$ suddenly came *before* t.

Pitts knew how to tackle the problem. He used modulo mathematics, which deals with numbers that circle back around on themselves like the hours of a clock. He showed McCulloch that the paradox of time $t + 1$ coming before time t wasn't a paradox at all, because in his calculations "before" and "after" lost their meaning. Time was removed from the equation altogether. If one were to see a lightning bolt flash on the sky, the eyes would send a signal to the brain, shuffling it through a chain of neurons. Starting with any given neuron in the chain, you could retrace the signal's steps and figure out just how long ago lightning struck. Unless, that is, the chain is a loop. In that case, the information encoding the lightning bolt just spins in circles, endlessly. It bears no connection to the time at which the lightning actually occurred. It becomes, as McCulloch put it, "an idea wrenched out of time." In other words, a memory.

By the time Pitts finished calculating, he and McCulloch had on their hands a mechanistic model of the mind, the first application of computation to the brain, and the first argument that the brain, at bottom, is an information processor. By stringing simple binary neurons into chains and loops, they had shown that the brain could implement every possible logical operation and compute anything that could be computed by one of Turing's hypothetical machines. Thanks to those ouroboric loops, they had also found a way for the brain to abstract a piece of information, hang on to it, and abstract it yet again, creating rich, elaborate hierarchies of lingering ideas in a process we call "thinking."

McCulloch and Pitts wrote up their findings in a now-seminal paper, "A Logical Calculus of Ideas Immanent in Nervous Activity," published in the *Bulletin of Mathematical Biophysics.* Their model was vastly oversimplified for a biological brain, but it succeeded at showing a proof of principle. Thought, they said, need not be shrouded in Freudian mysticism or engaged in struggles between ego and id. "For the first time in the history of science," McCulloch announced to a group of philosophy students, "we know how we know."

Pitts had found in McCulloch everything he had needed—acceptance, friendship, his intellectual other half, the father he never had. Although he had only lived in Hinsdale for a short time, the runaway would refer to McCulloch's house as home for the rest of his life. For his part, McCulloch was just as enamored. In Pitts he had found a kindred spirit, his "bootlegged collaborator," and a mind with the technical prowess to bring McCulloch's half-formed notions to life. As he put it in a letter of reference about Pitts, "Would I had him with me always."[1]

Pitts was soon to make a similar impression on one of the towering intellectual figures of the 20th century, the mathematician, philosopher, and founder of cybernetics, Norbert Wiener. In 1943 Lettvin brought Pitts into Wiener's office at the Massachusetts Institute of Technology. Wiener didn't introduce himself or make small talk. He simply walked Pitts over to a blackboard where he was working out a mathematical proof. As Wiener worked, Pitts chimed in with questions and suggestions. According to Lettvin, by the time they reached the second blackboard, it was clear that Wiener had found his new right-hand man. Wiener would later write that Pitts was "without question the strongest young scientist whom I have ever met . . . I should be extremely astonished if he does not prove to be one of the two or three most important scientists of his generation, not merely in America but in the world at large."

So impressed was Wiener that he promised Pitts a PhD in mathematics at MIT, despite the fact that he had never graduated from high school—something that the strict rules at the University of Chicago prohibited. It was an offer Pitts couldn't refuse. By the fall of 1943 Pitts had moved into a Cambridge apartment, was enrolled as a special student at MIT, and was studying under one of the most influential scientists in the world. It was quite a long way from blue-collar Detroit.

Wiener wanted Pitts to make his model of the brain more realistic. Despite the leaps Pitts and McCulloch had made, their work had made barely a ripple among brain scientists—in part because the symbolic logic they'd employed was hard to decipher, but also because their stark and oversimplified model didn't capture the full messiness of the biological brain. Wiener, however, understood the implications of what they'd done, and knew that a more realistic model would be game-changing. He also realized that it ought to be possible for Pitts's neural networks to be implemented in manmade machines, ushering in his dream of a cybernetic revolution. Wiener figured that if Pitts was going to make a realistic model of the brain's 100 billion interconnected neurons, he was going to need statistics on his side. And statistics and probability theory were Wiener's area of expertise. After all, it had been Wiener who discovered a precise mathematical definition of information: the higher the probability, the higher the entropy and the lower the information content.

As Pitts began his work at MIT, he realized that although genetics must encode for gross neural features, there was no way our genes could predetermine the trillions of synaptic connections in the brain—the amount of information it would require was untenable. It must be the case, he figured, that we all start out with essentially random neural networks—highly probable states containing negligible information (a thesis that continues to be debated to the present day). He suspected that by altering the thresholds of neurons over time, randomness could give way to order and information could emerge. He set out to model the process using statistical mechanics. Wiener excitedly cheered him on, because he knew if such a model were embodied in a machine, that machine could *learn.*

"I now understand at once some seven-eighths of what Wiener says, which I am told is something of an achievement," Pitts wrote in a letter to McCulloch in December of 1943, some three months after he'd arrived. His work with Wiener was "to constitute the first adequate discussion of statistical mechanics, understood in the most general possible sense, so that it includes for example the problem of deriving the psychological, or statistical, laws of behavior from the microscopic laws of neurophysiology . . . Doesn't it sound fine?"

That winter, Wiener brought Pitts to a conference he organized in Princeton with the mathematician and physicist John von Neumann, who was equally impressed with Pitts's mind. Thus formed the beginnings of the group who would become known as the cyberneticians, with Wiener, Pitts, McCulloch, Lettvin, and von Neumann its core. And among this rarefied group, the formerly homeless runaway stood out. "None of us would think of publishing a paper without his corrections and approval," McCulloch wrote. "[Pitts] was in no uncertain terms the genius of our group," said Lettvin. "He was absolutely incomparable in the scholarship of chemistry, physics, of everything you could talk about history, botany, etc. When you asked him a question, you would get back a whole textbook . . . To him, the world was connected in a very complex and wonderful fashion."[2]

The following June, 1945, von Neumann penned what would become a historic document entitled *First Draft of a Report on the EDVAC,* the first published description of a stored-program binary computing machine—the modern computer. The EDVAC's predecessor, the ENIAC, which took up 1,800 square feet of space in Philadelphia, was more like a giant electronic calculator than a computer. It was possible to reprogram the thing, but it took several operators several weeks to reroute all the wires and switches to do it. Von Neumann realized that it might not be necessary to rewire the machine every time you wanted it to perform a new function. If you could take each configuration of the switches and wires, abstract them, and encode them symbolically as pure information, you could feed them into the computer the same way you'd feed it data, only now the data would include the very programs that manipulate the data. Without having to rewire a thing, you'd have a universal Turing machine.

To accomplish this, von Neumann suggested modeling the computer after Pitts and McCulloch's neural networks. In place of neurons, he suggested vacuum tubes, which would serve as logic gates, and by stringing them together exactly as Pitts and McCulloch had discovered, you could carry out any computation. To store the programs as data, the computer would need something new: a memory. That's where Pitts's loops came into play. "An element which stimulates itself will hold a stimulus indefinitely," von Neumann wrote in his report, echoing Pitts and employing his modulo mathematics. He detailed every aspect of this new computational architecture. In the entire report, he cited only a single paper: "A Logical Calculus" by McCulloch and Pitts.

By 1946 Pitts was living on Beacon Street in Boston with Oliver Selfridge, an MIT student who would become "the father of machine perception"; Hyman Minsky, the future economist; and Lettvin. He was teaching mathematical logic at MIT and working with Wiener on the statistical mechanics of the brain. The following year, at the Second Cybernetic Conference, Pitts announced that he was writing his doctoral dissertation on probabilistic three-dimensional neural networks. The scientists in the room were floored. "Ambitious" was hardly the word to describe the mathematical skill that it would take to pull off such a feat. And yet, everyone who knew Pitts was sure that he could do it. They would be waiting with bated breath.

In a letter to the philosopher Rudolf Carnap, McCulloch cataloged Pitts's achievements. "He is the most omniverous of scientists and scholars. He has become an excellent dye chemist, a good mammalogist, he knows the sedges, mushrooms and the birds of New England. He knows neuroanatomy and neurophysiology from their original sources in Greek, Latin, Italian, Spanish, Portuguese, and German for he learns any language he needs as soon as he needs it. Things like electrical circuit theory and the practical soldering in of power, lighting, and radio circuits he does himself. In my long life, I have never seen a man so erudite or so really practical." Even the media took notice. In June 1954 *Fortune* magazine ran an article featuring the 20 most talented scientists under 40; Pitts was featured, next to Claude Shannon and James Watson. Against all odds, Walter Pitts had skyrocketed into scientific stardom.

Some years earlier, in a letter to McCulloch, Pitts wrote, "About once a week now I become violently homesick to talk all evening and all night to you." Despite his success, Pitts had become homesick—and home meant McCulloch. He was coming to believe that if he could work with McCulloch again, he would be happier, more productive, and more likely to break new ground. McCulloch, too, seemed to be floundering without his bootlegged collaborator.

Suddenly, the clouds broke. In 1952 Jerry Wiesner, associate director of MIT's Research Laboratory of Electronics, invited McCulloch to head a new project on brain science at MIT. McCulloch jumped at the opportunity—because it meant he would be working with Pitts again. He traded his full professorship and his large Hinsdale home for a research associate title and a crappy apartment in Cambridge, and couldn't have been happier about it. The plan for the project was to use the full arsenal of information theory, neurophysiology, statistical mechanics, and computing machines to understand how the brain gives rise to the mind. Lettvin, along with the young neuroscientist Patrick Wall, joined McCulloch and Pitts at their new headquarters in Building 20 on Vassar Street. They posted a sign on the door: EXPERIMENTAL EPISTEMOLOGY.

With Pitts and McCulloch together again, and with Wiener and Lettvin in the mix, everything seemed poised for progress and revolution. Neuroscience, cybernetics, artificial intelligence, computer science—it was all on the brink of an intellectual explosion. The sky—or the mind—was the limit.

There was just one person who wasn't happy about the reunion: Wiener's wife. Margaret Wiener was, by all accounts, a controlling, conservative prude—and she despised McCulloch's influence on her husband. McCulloch hosted wild get-togethers at his family farm in Old Lyme, Connecticut, where ideas roamed free and everyone went skinny-dipping. It had been one thing when McCulloch was in Chicago, but now he was coming to Cambridge and Margaret wouldn't have it. And so she invented a story. She sat Wiener down and informed him that when their daughter, Barbara, had stayed at McCulloch's house in Chicago, several of "his boys" had seduced her. Wiener immediately sent an angry telegram to Wiesner: "Please inform [Pitts and Lettvin] that all connection between me and your projects is permanently abolished. They are your problem. Wiener." He never spoke to Pitts again. And he never told him why.[3]

For Pitts, this marked the beginning of the end. Wiener, who had taken on a fatherly role in his life, now abandoned him inexplicably. For Pitts, it wasn't merely a loss. It was something far worse than that: it defied logic.

And then there were the frogs. In the basement of Building 20 at MIT, along with a garbage can full of crickets, Lettvin kept a group of them. At the time, biologists believed that the eye was like a photographic plate that passively recorded dots of light and sent them, dot for dot, to the brain, which did the heavy lifting of interpretation. Lettvin decided to put the idea to the test, opening up the frogs' skulls and attaching electrodes to single fibers in their optic nerves.

Together with Pitts, McCulloch, and the Chilean biologist and philosopher Humberto Maturana, he subjected the frogs to various visual experiences—brightening and dimming the lights, showing them color photographs of their natural habitat, magnetically dangling artificial flies—and recorded what the eye measured before it sent the information off to the brain. To everyone's surprise, it didn't merely record what it saw, but filtered and analyzed information about visual features like contrast, curvature, and movement. "The eye speaks to the brain in a language already highly organized and interpreted," they reported in the now-seminal paper "What the Frog's Eye Tells the Frog's Brain," published in 1959.

The results shook Pitts's worldview to its core. Instead of the brain computing information digital neuron by digital neuron using the exacting implement of mathematical logic, messy, analog processes in the eye were doing at least part of the interpretive work. "It was apparent to him after we had done the frog's eye that even if logic played a part, it didn't play the important or central part that one would have expected," Lettvin said. "It disappointed him. He would never admit it, but it seemed to add to his despair at the loss of Wiener's friendship."

The spate of bad news aggravated a depressive streak that Pitts had been struggling with for years. "I have a kind of personal woe I should like your advice on," Pitts had written to McCulloch in one of his letters. "I have noticed in the last two or three years a growing tendency to a kind of melancholy apathy or depression. [Its] effect is to make the positive value seem to disappear from the world, so that nothing seems worth the effort of doing it, and whatever I do or what happens to me ceases to matter very greatly . . ."

In other words, Pitts was struggling with the very logic he had sought in life. Pitts wrote that his depression might be "common to all people with an excessively logical education who work in applied mathematics: It is a kind of pessimism resulting from an inability to believe in what people call the Principle of Induction, or the principle of the Uniformity of Nature. Since one cannot prove, or even render probable a priori, that the sun should rise tomorrow, we cannot really believe it shall."

Now, alienated from Wiener, Pitts's despair turned lethal. He began drinking heavily and pulled away from his friends. When he was offered his PhD, he refused to sign the paperwork. He set fire to his dissertation along with all of his notes and his papers. Years of work—important work that everyone in the community was eagerly awaiting—he burned it all, priceless information reduced to entropy and ash. Wiesner offered Lettvin increased support for the lab if he could recover any bits of the dissertation. But it was all gone.

Pitts remained employed by MIT, but this was little more than a technicality; he hardly spoke to anyone and would frequently disappear. "We'd go hunting for him night after night," Lettvin said. "Watching him destroy himself was a dreadful experience." In a way, Pitts was still 12 years old. He was still beaten, still a runaway, still hiding from the world in musty libraries. Only now his books took the shape of a bottle.

With McCulloch, Pitts had laid the foundations for cybernetics and artificial intelligence. They had steered psychiatry away from Freudian analysis and toward a mechanistic understanding of thought. They had shown that the brain computes and that mentation is the processing of information. In doing so, they had also shown how a machine could compute, providing the key inspiration for the architecture of modern computers. Thanks to their work, there was a moment in history when neuroscience, psychiatry, computer science, mathematical logic, and artificial intelligence were all one thing, following an idea first glimpsed by Leibniz—that man, machine, number, and mind all use information as a universal currency. What appeared on the surface to be very different ingredients of the world—hunks of metal, lumps of gray matter, scratches of ink on a page—were profoundly interchangeable.

There was a catch, though: this symbolic abstraction made the world transparent but the brain opaque. Once everything had been reduced to information governed by logic, the actual mechanics ceased to matter—the tradeoff for universal computation was ontology. Von Neumann was the first to see the problem. He expressed his concern to Wiener in a letter that anticipated the coming split between artificial intelligence on one side and neuroscience on the other. "After the great positive contribution of Turing-cum-Pitts-and-McCulloch is assimilated," he wrote, "the situation is rather worse than better than before. Indeed these authors have demonstrated in absolute and hopeless generality that anything and everything . . . can be done by an appropriate mechanism, and specifically by a neural mechanism—and that even one, definite mechanism can be 'universal.' Inverting the argument: Nothing that we may know or learn about the functioning of the organism can give, without 'microscopic,' cytological work any clues regarding the further details of the neural mechanism."

This universality made it impossible for Pitts to provide a model of the brain that was practical, and so his work was dismissed and more or less forgotten by the community of scientists working on the brain. What's more, the experiment with the frogs had shown that a purely logical, purely brain-centered vision of thought had its limits. Nature had chosen the messiness of life over the austerity of logic, a choice Pitts likely could not comprehend. He had no way of knowing that while his ideas about the biological brain were not panning out, they were setting in motion the age of digital computing, the neural network approach to machine learning, and the so-called connectionist philosophy of mind. In his own mind, he had been defeated.

On Saturday, April 21, 1969, his hand shaking with an alcoholic's delirium tremens, Pitts sent a letter from his room at Beth Israel Hospital in Boston to McCulloch's room down the road at the Cardiac Intensive Care Ward at Peter Bent Brigham Hospital. "I understand you had a light coronary; . . . that you are attached to many sensors connected to panels and alarms continuously monitored by a nurse, and cannot in consequence turn over in bed. No doubt this is cybernetical. But it all makes me most abominably sad." Pitts himself had been in the hospital for three weeks, having been admitted with liver problems and jaundice. On May 14, 1969, Walter Pitts died alone in a boarding house in Cambridge, of bleeding esophageal varices, a condition associated with cirrhosis of the liver. Four months later, McCulloch passed away, as if the existence of one without the other were simply illogical, a reverberating loop wrenched open.

Notes

1. All letters retrieved from American Philosophical Society, *Warren S. McCulloch Papers*, BM139, Series I: Correspondence 1931–1968, Folder "Pitts, Walter."
2. All Jerome Lettvin quotes taken from James A. Anderson and Edward Rosenfield, eds., *Talking Nets: An Oral History of Neural Networks* (Cambridge, MA: MIT Press, 2000).
3. Flo Conway and Jim Siegelman, *Dark Hero of the Information Age: In Search of Norbert Wiener, the Father of Cybernetics* (New York: Basic Books, 2006).

ROSE GEORGE

A Very Naughty Little Girl

FROM *Longreads*

SHE WAS A NAME on a plaque and a face on a wall. I ate beneath her portrait for three years and paid it little attention except to notice that the artist had made her look square. There were other portraits of women to hold my attention on the walls of Somerville, my Oxford college: Indira Gandhi, who left without a degree, and Dorothy Hodgkin, a Nobel Prize–winner in chemistry. In a room where we had our French-language classes, behind glass that was rumored to be bulletproof, there was also a bust of Margaret Thatcher, a former chemistry undergraduate. Somerville was one of only two women's colleges of the University of Oxford while I was there, from 1988 to 1992, and the walls were crowded with strong, notable women. (The college has since gone co-ed.)

The plaque saying VAUGHAN was on the exterior wall of my first-year student residence, a building named after Vaughan, Dame Janet Maria, the woman in the portrait and principal of Somerville between 1945 and 1967. She was still alive when I was an undergraduate, and, according to her obituaries, was known for driving around Oxford in a yellow Mini; for always dressing in tweeds; and for going to the Bodleian Library even in her late 80s and inadvertently annoying other readers when her hearing aid hummed and whistled. But when I arrived at Somerville, and was assigned to Vaughan, I thought only with some relief that everyone would finally be able to spell my Welsh third name, usually a puzzle even to English speakers. I did not think back to my two surgeries, or to my birth, where bags of someone else's blood and plasma would have hung from hooks and saved my life, and thank

Janet Vaughan for her role in helping to make that standard medical practice. But I should have.

Blood. We all have it, this liquid that is nearly half water, red in color, that "circulates in the arteries and veins, carrying oxygen to and carbon dioxide from the tissues of the body," an *Oxford Dictionary* definition. Blood is common, ubiquitous, inevitable. But it is also so much more than its dictionary description. It is more expensive than oil. Every two seconds, someone needs some. Every day, millions of people receive blood from anonymous strangers in a wondrous procedure that has become banal.

In the United Kingdom, where I live, the system of widespread donation of blood by anonymous volunteers, and its transfusion into people who need it, dates back not even a century. The National Blood Service began as the Blood Transfusion Service in 1946; the National Health Service was founded in 1948. But humans have been curious about blood probably since the first human bled. The first date on the history timeline of the National Blood Service is 1628, when William Harvey demonstrated that blood circulates around the body, but the practice of bloodletting is usually dated back to Egyptians of 3,000 years ago. Letting out blood eased the imbalance of the four humors—blood, phlegm, black bile, yellow bile—that governed the body.

Transfusion—the transfer of blood from one creature to another—is also ancient. Romans, wrote Pliny the Elder, ran to drink the blood of dying or dead gladiators, to gain some of their strength and force. Blood was also thought to carry personality, so when Jean-Baptiste Denis, doctor to Louis XIV, treated a feverish 16-year-old patient with a blood transfusion in 1667, he thought the "mild and laudable" blood of a lamb his best bet. The patient recovered, though a madman treated with calf's blood did not. Transfusion was largely avoided for 150 years. The Victorians tried, but humans given human blood kept dying, and only when Karl Landsteiner discovered blood types in 1909 (and that some mixed fatally) did transfusion become acceptable.

Transfusion is now unremarkable. A car-accident victim can require 100 pints of blood. Coronary bypass surgery can require blood from 20 donors, while a premature baby can be saved with three teaspoonsful. In the United States 15.7 million pints of blood are donated every year, and more than 100 million globally, according to the World Health Organization. But in 1920s Oxford, when Janet Vaughan was a medical sciences undergraduate at Somerville, the speedy transfer of blood from one human to another en masse was unthinkable.

Vaughan was born to privilege. Her father was a headmaster of fine public schools such as Rugby. Her mother, Madge Symonds, was a beauty, though Vaughan described her as "a caged butterfly or hummingbird," and the cage was her life as a headmaster's wife in stuffy schools where she ate chicken bones with her fingers to shock the butler. Madge was great friends with Virginia Woolf, Janet Vaughan's second cousin, and the character of Sally Seton in *Mrs. Dalloway* is based on her. The Vaughans were connected, but not wealthy. Janet was given indifferent schooling, including a headmistress who described her as "too stupid to be educated." She ignored that, read voraciously, and after three attempts, was accepted at Somerville to study medical sciences. She arrived, she said, with nothing more than "a little ladylike botany," yet graduated with a first-class degree, and set about being a physician. For her obstetrics rotation at University College Hospital, London, she was sent into London's slums. "Terrible poverty," she told the journalist Polly Toynbee, when Vaughan was interviewed as one of six women chosen by the BBC to be *Women of Our Century*. She encountered "a woman with no bed except newspaper. A husband who said, 'Ain't there no male doctors? I'd rather have a black man than you.'" She saw lines of children sitting up in bed with rheumatic hearts, who would die. She saw that poverty is deadly. "How anyone could do medicine in those days and not become a socialist I find hard to understand," Vaughan wrote. "What I hated most was people's acceptance: 'Yes, I have had seven children and buried six, it was God's will.' I hated God's will with a burning hatred."

The slums introduced her to real poverty, but also to blood, her lifelong interest. Anemia was endemic, and by the time she qualified, she began to wonder why the standard cure for anemia was arsenic. She was a pathologist by now: her mother had died, and she thought her widowed father needed her. Pathology, with a more stable routine than doctoring, would make her available for him. But a pathologist can still read. She was trained at Oxford, and at Oxford you are trained to read ferociously. So she had read of the work of Dr. George Minot, an American physician who had been treating pernicious anemia with raw liver extract. Vaughan thought this made more sense than arsenic, so she approached her professor of medicine, a man who did not see a woman who was young and think both those things to be handicaps. "I went to the professor," she told the BBC, "and said, can we test it on a patient?" He said no, but she could try it first on a dog, if she produced the extract herself. He gave her some money, and off she went to collect as many pails and mincing machines as she could. She did the rounds of her friends. She borrowed Virginia Woolf's mincer, and minced and minced. It became a scene in Woolf's *A Room of One's Own*, Janet with her mincers, Minot's paper propped on the kitchen table, a parody of expected domesticity.

The extract was fed to two dogs, who sickened. Janet said, no more dogs, and took the extract herself. "The next morning when I came back to the hospital there were all the professors of medicine, chemistry, surgery, waiting on the doorstep to see if I was still alive." It was fed to a patient, "a nice old labouring man." The patient survived, and of course a senior professor took all the credit for the miraculous new treatment. Janet didn't much mind: she had other things to do. Her father had remarried, and she could travel, so she got a Rockefeller Scholarship to Harvard, where there were no women. She wasn't allowed to work with patients and she wasn't allowed to work with mice either, for when she ordered some Boston mice, a famous lab variety, she was told there were none to spare. No matter. She sourced some "excellent Philadelphian mice," but Harvard didn't allow them. She ended up with pigeons, using them to do groundbreaking research on vitamin B_{12} in blood that wasn't fully acknowledged for 50 years. She called them her Bloody Pigeons.

How I love the brisk nervelessness of this woman. Some of it comes from privilege. But much of it is her own, as much as her fictional room was. She had the confidence to make fissures in patriarchal concrete, but also the confidence to get married, because she wanted to. With her husband, David Gourlay, she moved into Gordon Square in Bloomsbury, and Vaughan went to work at a hospital where no one spoke to her and physicians asked for her advice—by then her reputation was significant—only by letter. She treated more anemic patients with liver, though the patients then said, "Don't give me any more of that medicine, doctor. It makes me hungry and I can't afford [to eat]." She taught her patients to fight authorities to get extra milk, for the extra iron it would give them. She taught her students that to practice medicine, they must learn to deal with the public assistance board as well as the hospital dispensary. She had two daughters. She was busy, and happy.

But war was coming. First, it came to Spain. Her Bloomsbury friends went to fight, and Vanessa Bell's son Julian was killed. Vaughan began to work with the Committee for Spanish Medical Aid and joined the Communist Party, though she soon lapsed. She said that no one seemed to notice. She sold possessions to raise money for Basque children; she spoke on soapboxes at street corners. She welcomed to London Dr. Duran-Jordá, an exceptionally gifted Spanish hematologist who had worked out how to store large supplies of blood so it could be used in a war situation. By now, taking blood from one human and putting it in another was understood, but storing that blood so it didn't spoil was not. Vaughan read that Russians had also stored blood, taken from fatal road accidents and kept at low temperatures, then used for civilian needs. She noted all this, and she kept it safe until 1938, when the Munich Agreement was signed, allowing Nazi Germany to annex Sudetenland.

We think now, in our era of wars that only happen elsewhere, of 1938 as safely prewar. But it can't have felt like it. A bombing blitz on London was seriously expected to follow Munich. Someone came to Hammersmith and told the medical school to be ready for 57,000 casualties in London that weekend. And Vaughan realized immediately that casualties would need blood. They would need a lot of blood.

In the Wellcome Library in London, I find a propaganda film published by the Ministry of Information in the 1940s, a time when neither "propaganda" nor "Ministry of Information" sounded sinister. The film is called *Blood Transfusion,* and is narrated by accents that now sound cut-glass and royal, but then were normal. It tells us that blood transfusion was widely used in World War I on the Western Front, including by Dr. Oswald Robinson of Toronto. Donors were easily on hand, and by then it had been understood that adding sodium citrate to blood stopped coagulation and made blood easier to store.

The film moves to 1921, in Camberwell, where we meet Mr.

Percy Oliver, leader of a local branch of the British Red Cross Society. In 1921 there was no mass storage of blood, nor any organized recruitment of blood donors. Blood donation was done, but ad hoc. The technical term was "on the hoof." Oliver's actions helped to change that, when the telephone rang on an October evening in the south London suburb of Camberwell, and someone from King's College Hospital asked for urgent blood donors. Oliver volunteered, and so did all the nurses on the committee. From this one call in 1921, Oliver set up a network of blood donors, an office with donors' details on index cards and a telephone that was always manned, and he called it the London Blood Service. Donors were sought and their blood used immediately. The service dealt with 737 calls in 1926; 2,442 in 1931; and by 1938 nearly 7,000. Oliver's model was copied and spread, so that by the war, most parts of the country could reach a donor panel. It was an amazing achievement from a man who still worked full-time at the local council, and from each donor who gave some of their eight or nine pints of blood, freely and for no reward.

But in a war, it would not do.

By 1938 science knew how to transfuse blood and store it temporarily. But Janet Vaughan knew that on the hoof would not be possible in wartime. She needed to work out how to store large quantities of blood, safely, and transport it, safely. Blood supply needed to go from Mom and Pop to Walmart. Yet there was resistance. Douglas Starr, in his 1998 book *Blood,* quotes the British secretary of war, asked in 1937 what the nation proposed to do about a mass blood supply. The secretary was dismissive. Blood could not be stored for long or in great quantities, he said. On the hoof was better. "It was more satisfactory to store our blood in our people."

Janet Vaughan did not agree, and Janet Vaughan did something about it. Her medical director gave her £100, and she sent off her assistants in taxis to find all the tubing that London shops could provide. They made up crude transfusion sets, and set about bleeding. (I am struck by this word, which Vaughan and contemporary accounts use routinely. It has been retired in the modern blood-supply service. Today, I receive letters asking me to "donate" or "give blood," but not to be bled, or to sit on a "bleeding couch." Is it too brutal? Does the word make us too animal? Now the emphasis is on altruism, not the baseness of a body fluid.)

The bombs did not fall, and Vaughan's team diverted the blood collected at Hammersmith to hospital use. But war was going to come, so Janet began planning. The minutes of the meetings of the Emergency Blood Transfusion Service exist in the archives of the Wellcome Library, and they are as rich as blood. The service was actually a gathering of Vaughan's peers from various hospitals, held in the Gourlays' Bloomsbury flat. The meetings were always in the evening, after the day's work was done, and they lasted hours. I picture these meetings. Janet Vaughan would be wearing a tweed suit. She would be kind but brisk. Later, someone described her as "down to earth but like air on a mountain." The other doctors might have bow ties. They would smoke pipes. They would drink tea, or gin and tonics, and eat crumpets. They would do this while deciding on the size of bottles, or what kind of armrests to put on the "bleeding chairs," and they would change modern medicine.

London needed blood depots. The committee decided on four, one for each quarter of the city; two north of the river, two south. They would be near hospitals, but safely out of the city center, in case it was bombed. The service needed money, so they worked on a budget, which was doubled on the advice of Vaughan's boss. The precise methods of storing blood had to be decided upon: On April 5, 1939, the first meeting, the minutes record a suggestion that 50 cc of 3.8 percent citrate containing 0.1 percent glucose should be added to every 450 cc of blood. Donors, they thought at first, would be group O only, the universal blood type that can be safely transfused into most people, and there should be 9,000 of them per depot. Other issues arose. Of course donors must be screened, as much as circumstances permitted. Syphilis was a worry, particularly as promiscuity always rises in wartime. The practicality of testing all donations for syphilis was one issue; whether or not to tell unwitting carriers that they had been infected by their spouses was another.

The bottles, though. Among all calculations detailed painstakingly in the committee minutes over the months—8 million Londoners, therefore 2 million people per depot, 65,000 estimated daily casualties, therefore 10,000 per depot; and from wanting only O to accepting all blood types—the bottles were vital. Without the proper storage vessel, all the sodium citrate in the world wouldn't keep blood safe, or make it portable. The choices were few but they were tricky: A Beattie waisted type? A whiskey cap on a United Dairies bottle? A McCartney screw cap on a United

Dairies bottle? Or a modified McCartney bottle of the LCC type? "The children complained that the flat was littered with old bottles," wrote Vaughan in her unpublished autobiography "Jogging Along."

Through the summer the deliberations continued. The committee was an informal one, but then Vaughan sent a memorandum about their plans to Professor Topley of the London School of Hygiene, who was known to be organizing emergency services. Her boss heard of it and called her "a very naughty little girl." As if that would stop her.

By the second week of June, they had decided to use Wall's ice-cream vans to transport blood. They would call up registered donors by using runner boys. The bottle would be a modified milk bottle with a narrow waist and an aluminum screw cap lined with a soft rubber disk. This became known as the MRC bottle, after the committee was taken over by the Medical Research Council.

Janet was to run the northern depot, in Slough. She set off alone to see the medical officer of health (MOH), to find premises. "How fortunate I was to go to Slough where everyone—mad as they thought me at the time—was more than willing to help me." There was an unshackled energy about Slough that appealed to this child of boarding school, Oxford, and Bloomsbury. She thought it "a frontier town," grown up after World War I around a vast trading estate, full of migrant workers "with no settled traditions and customs to be disturbed." She was directed to Noel Mobbs, chairman of the Slough Trading Estate and of a social center. Mr Mobbs did not believe a war was coming, but he said the depot could move into the social center, that there was space for cold rooms to be built. There was also a bar.

Agreements were signed. Hitler prepared to invade Poland. Three days before Britain declared war, on September 3, 1939, Janet Vaughan received a telegram from the Medical Research Council. It read, "Start bleeding."

Slough Depot. Wartime. The place bustled with nurses, secretaries, telephonists, medical technicians, drivers. There were 100 staff, including 2 local girls to make all the sterile plasma and serum, "and sterile they kept in spite of the really unsuitable dusty conditions in which they had to work." And the drivers who must drive the Wall's ice-cream vans full of blood. These drivers: I pictured them as forthright young women in heavy coats, like the forthright young women in heavy coats in Sarah Waters's *The Night Watch.* But I was wrong, because in Slough, there was "Mrs. E. O. Franklin's chauffeur Brady, a mad Irishman," who kept the vans ticking over. There were voluntary drivers, and unexpected drivers who in times of dire need, such as when Liverpool was bombed and Liverpool had no blood, were recruited from the bar. Drinkers became drivers. Janet loved this bar, for what it meant to her staff: "My young drivers, girls, coming in late at night having driven through terrible weather and blackout, to be able to get some whisky in the war was very important."

They were not all young girls. One regular driver was Lady Dunstan, "who must have been at least 70. She always wore a string of pearls and a toque rather like Queen Mary, but she was never daunted." An elegant pensioner, driving an ice-cream van around bombed streets: I couldn't have invented it. Another driver was "a remarkable old lady whose only interest in life before the war had been her string of ponies and her bridge. She came and said she wanted a job and we set her down amongst the young technicians to fix a singular nasty wire filter that was being used at that time for stored plasma." When she was ill, as she was occasionally, "she used to send her chauffeur in the Rolls Royce to fetch her a supply of filters to fit at home in bed. One of her friends said she had never been so happy in her life before. She knew we depended on her work, as we did, and through us casualties all over the country."

And who sat on the bleeding couches and gave their blood? Everyone. Some were recruited in factories or army units or offices, where large groups of people could be appealed to easily. "But the housewife in the country village and small town was often a most faithful and regular donor." Many gave every three months throughout the war, "feeling it the one personal contribution they could make to the war effort." Richard Titmuss, in *The Gift Relationship,* reported the results of a survey into why people donate blood. One man wrote, "1941. War. Blood needed. I had some. Why not?"

The bleeding of donors was undertaken in the depot and by mobile teams sent out into the surrounding small towns and villages, where they set up a temporary bleeding center in a town hall, factory restroom, church hall, or village public house. Janet was very proud of the quality and sterility of "the bleed." At first, blood was requested by telephone. But "they soon learnt that Slough could hear and see the bombs falling and would arrive." The bombs began to drop in the fall of 1940. The ice-cream vans would get near the bomb sites, and deliver their blood to the hospital where the casualties were being taken. This experience created medical innovation: they wore electric lights on their foreheads, like miners, so that if lights were off and blackout was in operation, they could still see where to stick in their needles.

They changed the needle design because of one incident with a burned girl on the Great West Road. Vaughan arrived and found a little girl, horrifically burned after she had taken alight from the house electric fire. She left the girl to die, because she had to see who she could save and who she couldn't, but after saving who she could she returned to the girl and found her still alive. Her legs and arms were so burned, she had no veins. And Janet again remembered something she had read, that you could give blood into bones. "That was the great thing about medicine in the war, you could take risks because people died so they were no worse off if they died because of what you did." She took the biggest needle she could find and stuck it in the girl's breastbone, and told a nurse to pump in blood. (When depot staff couldn't find a vein, they called it "digging for victory.") Two hours later, the nurse had got two pints in. The girl lived, and applied to Somerville years later, when Vaughan was principal. Harriet Proudfoot, that burned little girl, was given the choice of three Oxford colleges to choose from, but remembered Vaughan, who visited her and sucked blood out of her undamaged ear. And all the while "she explained what she was doing and spoke to me as if I was as interested and intelligent as she was." When Harriet applied to Oxford, the only college she put on her form was Somerville. "So," said Janet Vaughan of this, "nice things happen."

The depots treated thousands of casualties, but the staff also did scientific research. They learned that theories about how to treat shock were wrong: rather than having the expected low blood pressure, shock victims had high blood pressure, and needed large volumes of blood. Each trauma victim received on average two and a half milk bottles of blood, with two bottles of blood used for one bottle of plasma. Sometimes, they quickly learned, dried serum and no blood was adequate treatment. This was essential for the mass evacuation of Dunkirk in the summer of 1940, when the depot sent all the blood it had to the coast. But the casualties kept coming, and the system could not cope, even though the Americans by now were sending over blood and plasma on ships, under the Blood for Britain program. Vaughan and her staff had also been working on using plasma, but the plasma looked cloudy and full of clots, so they hadn't dared risk it until Dunkirk. "We knew men must die if we didn't transfuse them, so we took a risk on our very odd-looking plasma." It was another risk justified by war, and the plasma worked "like magic."

By the spring of 1941, when the air raids stopped, they thought the demand for blood would drop, but it didn't. "Surgeons and physicians had learnt to appreciate the value of transfusion as a therapeutic aid. Throughout the war there was a steady increase of transfusion practice throughout the country. In many cases no doubt the pendulum swung too far and unnecessary transfusions were given, but on the whole the educative value of the war time transfusion service was great." The war, and the depots of London —along with army depots—taught us the value of blood.

Through the war, then, delivering blood. Vaughan learned to say yes to any request, because "what men and women need in a desperate emergency is reassurance. They can hold on if help is coming, and—given the lead—other men and women will always be prepared to give that help." Just before D-day, Janet received a phone call from the head of Emergency Medical Services. "Janet, we have made no arrangements for the Ports, will you look after them?" She said, yes, of course, having no idea what looking after the Ports would entail. "As so often we heard no more, but I can only hope that the Ports received reassuring messages that Slough would come if needed."

The war was ending, and although Vaughan had had five years of death and burns and bombs, when she was asked to go to the Nazi death camp Belsen to research how best to feed starving people, she said yes. She said yes, for science, because people who had been as starved and annihilated as the inmates of Belsen needed emergency nutrition, and the prevailing medical thought was that this should consist of hydrolysates, a strong protein in liquid form. She was driven over the Rhine on wooden pontoon bridges, and she waved to troops returning from the front. She saw hundreds of forced laborers in striped pajamas, spat out from their camps and wandering over the countryside. When she got home, she burned all her husband's striped pajamas.

They knew before they got there that they were near Belsen, because of the stench of shit and dead bodies. The senior officer thought they were there to help, but they said no, they were there to do research. That sounds brutal now, but Vaughan writes about it with no shame. She believed in science, and they would soon need to save all the prisoners of war who would emerge from Japanese camps. The science must be done, even if she found herself having to inject hydrolysates into skeletal men who saw her approaching with a needle and screamed *"Nicht crematorium!"* because the Nazis had sometimes injected the condemned with paraffin before sending them to the crematoriums. Vaughan writes that this was so that they burned better. This was horror, but Vaughan kept on, though she had to pick the living from the dying in piles of bodies; though she was attacked by naked, desperate men (and could only wave a bedpan at them). She did enough research to show that small amounts of food were a more efficient treatment of starvation than hydrolysates. She wrote a letter home on a scrap of paper, that said, "I am here—trying to do science in hell."

She did science in hell for a few weeks, then returned to England, and to a job as principal of Somerville College for the next 22 years. It was academia, but it was no sinecure: of course she rose at dawn to dictate all the correspondence needed to run an Oxford college, before setting off every day to put a full day of work in at her lab nearby. If callers to the college wondered where the principal was, she responded, "Do they think I sit knitting?" She became an expert on radiation, researching the metabolism of nuclear fission on humans and on the rabbits who were her test animals. A fellow scientist called her "our radioactive principal," which was more accurate than he expected, because if there was any risk that radiation had leaked, Janet Vaughan would disappear to have a bone biopsy from her tibia. She did this work for decades, and once answered the politician Shirley Williams's question as to why on earth she was handling plutonium at her age, with "what could be better than for someone in her seventies to do this work: I haven't long to live anyway." She fought to have women's colleges accepted as full Oxford colleges, and she increased the intake of science and medical undergraduates at Somerville. She never stopped being a socialist.

She lived long after retiring in 1967. She still wrote academic papers in her 80s. She was a dame, of course, a member of so many societies (my favorite: the Bone and Tooth Society), and loaded with honorary degrees. She was establishment, but yet not, to the end.

Polly Toynbee asked her in 1984 how she would like to be remembered. And this woman who was instrumental in setting up mass blood donation and transfusion; who dared to stick a large needle into the breastbone of a small burned girl; who did science in hell; who never stopped encouraging science in all ways, and women to do more of it, said, "As a scientist. That I have been able to solve, to throw light onto fascinating problems. But as a scientist who had a family. I don't want to be thought of as a scientist who just sat thinking. It's important you have a human life."

Dame Janet Maria Vaughan died in January 1993, aged 93. She had only recently stopped driving her Mini. By then, the Blood Transfusion Service, later the National Blood Service, then NHS Blood and Transplant, had existed for 47 years. Last year, it collected 1.9 million donations of blood. British hospitals use 8,000 units every day. Although some parts of blood still only last five days, it can now be separated, stored, and widely transported, partly because of a woman named Janet.

GABRIELLE GLASER

The False Gospel of Alcoholics Anonymous

FROM *The Atlantic*

J.G. IS A LAWYER in his early 30s. He's a fast talker and has the lean, sinewy build of a distance runner. His choice of profession seems preordained, as he speaks in fully formed paragraphs, his thoughts organized by topic sentences. He's also a worrier—a big one—who for years used alcohol to soothe his anxiety.

J.G. started drinking at 15, when he and a friend experimented in his parents' liquor cabinet. He favored gin and whiskey but drank whatever he thought his parents would miss the least. He discovered beer, too, and loved the earthy, bitter taste on his tongue when he took his first cold sip.

His drinking increased through college and into law school. He could, and occasionally did, pull back, going cold turkey for weeks at a time. But nothing quieted his anxious mind like booze, and when he didn't drink, he didn't sleep. After four or six weeks dry, he'd be back at the liquor store.

By the time he was a practicing defense attorney, J.G. (who asked to be identified only by his initials) sometimes drank almost a liter of Jameson in a day. He often started drinking after his first morning court appearance, and he says he would have loved to drink even more, had his schedule allowed it. He defended clients who had been charged with driving while intoxicated, and he bought his own Breathalyzer to avoid landing in court on drunk-driving charges himself.

In the spring of 2012 J.G. decided to seek help. He lived in

Minnesota—the Land of 10,000 Rehabs, people there like to say—and he knew what to do: check himself into a facility. He spent a month at a center where the treatment consisted of little more than attending Alcoholics Anonymous meetings. He tried to dedicate himself to the program even though, as an atheist, he was put off by the faith-based approach of the 12 steps, 5 of which mention God. Everyone there warned him that he had a chronic, progressive disease and that if he listened to the cunning internal whisper promising that he could have just one drink, he would be off on a bender.

J.G. says it was this message—that there were no small missteps, and 1 drink might as well be 100—that set him on a cycle of bingeing and abstinence. He went back to rehab once more and later sought help at an outpatient center. Each time he got sober, he'd spend months white-knuckling his days in court and his nights at home. Evening would fall and his heart would race as he thought ahead to another sleepless night. "So I'd have one drink," he says, "and the first thing on my mind was: *I feel better now, but I'm screwed. I'm going right back to where I was. I might as well drink as much as I possibly can for the next three days.*"

He felt utterly defeated. And according to AA doctrine, the failure was his alone. When the 12 steps don't work for someone like J.G., Alcoholics Anonymous says that person must be deeply flawed. The Big Book, AA's bible, states:

> Rarely have we seen a person fail who has thoroughly followed our path. Those who do not recover are people who cannot or will not completely give themselves to this simple program, usually men and women who are constitutionally incapable of being honest with themselves. There are such unfortunates. They are not at fault; they seem to have been born that way.

J.G.'s despair was only heightened by his seeming lack of options. "Every person I spoke with told me there was no other way," he says.

The 12 steps are so deeply ingrained in the United States that many people, including doctors and therapists, believe attending meetings, earning one's sobriety chips, and never taking another sip of alcohol is the only way to get better. Hospitals, outpatient clinics, and rehab centers use the 12 steps as the basis for treatment. But although few people seem to realize it, there are alternatives, including prescription drugs and therapies that aim to help patients learn to drink in moderation. Unlike Alcoholics Anonymous, these methods are based on modern science and have been proved, in randomized, controlled studies, to work.

For J.G., it took years of trying to "work the program," pulling himself back onto the wagon only to fall off again, before he finally realized that Alcoholics Anonymous was not his only, or even his best, hope for recovery. But in a sense, he was lucky: many others never make that discovery at all.

The debate over the efficacy of 12-step programs has been quietly bubbling for decades among addiction specialists. But it has taken on new urgency with the passage of the Affordable Care Act, which requires all insurers and state Medicaid programs to pay for alcohol- and substance-abuse treatment, extending coverage to 32 million Americans who did not previously have it and providing a higher level of coverage for an additional 30 million.

Nowhere in the field of medicine is treatment less grounded in modern science. A 2012 report by the National Center on Addiction and Substance Abuse at Columbia University compared the current state of addiction medicine to general medicine in the early 1900s, when quacks worked alongside graduates of leading medical schools. The American Medical Association estimates that out of nearly 1 million doctors in the United States, only 582 identify themselves as addiction specialists. (The Columbia report notes that there may be additional doctors who have a subspecialty in addiction.) Most treatment providers carry the credential of addiction counselor or substance-abuse counselor, for which many states require little more than a high-school diploma or a GED. Many counselors are in recovery themselves. The report stated: "The vast majority of people in need of addiction treatment do not receive anything that approximates evidence-based care."

Alcoholics Anonymous was established in 1935, when knowledge of the brain was in its infancy. It offers a single path to recovery: lifelong abstinence from alcohol. The program instructs members to surrender their ego, accept that they are "powerless" over booze, make amends to those they've wronged, and pray.

Alcoholics Anonymous is famously difficult to study. By necessity, it keeps no records of who attends meetings; members come and go and are, of course, anonymous. No conclusive data exist on how well it works. In 2006 the Cochrane Collaboration, a health-care research group, reviewed studies going back to the 1960s and found that "no experimental studies unequivocally demonstrated the effectiveness of AA or [12-step] approaches for reducing alcohol dependence or problems."

The Big Book includes an assertion first made in the second edition, which was published in 1955: that AA has worked for 75 percent of people who have gone to meetings and "really tried." It says that 50 percent got sober right away, and another 25 percent struggled for a while but eventually recovered. According to AA, these figures are based on members' experiences.

In his recent book, *The Sober Truth: Debunking the Bad Science Behind 12-Step Programs and the Rehab Industry,* Lance Dodes, a retired psychiatry professor from Harvard Medical School, looked at Alcoholics Anonymous's retention rates along with studies on sobriety and rates of active involvement (attending meetings regularly and working the program) among AA members. Based on these data, he put AA's actual success rate somewhere between 5 and 8 percent. That is just a rough estimate, but it's the most precise one I've been able to find.

I spent three years researching a book about women and alcohol, *Her Best-Kept Secret: Why Women Drink—and How They Can Regain Control,* which was published in 2013. During that time, I encountered disbelief from doctors and psychiatrists every time I mentioned that the Alcoholics Anonymous success rate appears to hover in the single digits. We've grown so accustomed to testimonials from those who say AA saved their life that we take the program's efficacy as an article of faith. Rarely do we hear from those for whom 12-step treatment doesn't work. But think about it: How many celebrities can you name who bounced in and out of rehab without ever getting better? Why do we assume they failed the program, rather than that the program failed them?

When my book came out, dozens of Alcoholics Anonymous members said that because I had challenged AA's claim of a 75 percent success rate, I would hurt or even kill people by discouraging attendance at meetings. A few insisted that I must be an "alcoholic in denial." But most of the people I heard from were desperate to tell me about their experiences in the American treatment industry. Amy Lee Coy, the author of the memoir *From Death Do I Part: How I Freed Myself from Addiction,* told me about her eight trips to rehab, starting at age 13. "It's like getting the same antibiotic for a resistant infection—eight times," she told me. "Does that make sense?"

She and countless others had put their faith in a system they had been led to believe was effective—even though finding treatment centers' success rates is next to impossible: facilities rarely publish their data or even track their patients after discharging them. "Many will tell you that those who complete the program have a 'great success rate,' meaning that most are abstaining from drugs and alcohol while enrolled there," says Bankole Johnson, an alcohol researcher and the chair of the psychiatry department at the University of Maryland School of Medicine. "Well, no kidding."

Alcoholics Anonymous has more than 2 million members worldwide, and the structure and support it offers have helped many people. But it is not enough for everyone. The history of AA is the story of how one approach to treatment took root before other options existed, inscribing itself on the national consciousness and crowding out dozens of newer methods that have since been shown to work better.

A meticulous analysis of treatments, published more than a decade ago in *Handbook of Alcoholism Treatment Approaches* but still considered one of the most comprehensive comparisons, ranks AA 38th out of 48 methods. At the top of the list are brief interventions by a medical professional; motivational enhancement, a form of counseling that aims to help people see the need to change; and acamprosate, a drug that eases cravings. (An oft-cited 1996 study found 12-step facilitation—a form of individual therapy that aims to get the patient to attend AA meetings—as effective as cognitive-behavioral therapy and motivational interviewing. But that study, called Project Match, was widely criticized for scientific failings, including the lack of a control group.)

As an organization, Alcoholics Anonymous has no real central authority—each AA meeting functions more or less autonomously—and it declines to take positions on issues beyond the scope of the 12 steps. (When I asked to speak with someone from the General Service Office, AA's administrative headquarters, regarding AA's stance on other treatment methods, I received an email stating: "Alcoholics Anonymous neither endorses nor opposes other approaches, and we cooperate widely with the medical profession." The office also declined to comment on whether AA's efficacy has been proved.) But many in AA and the rehab industry insist the 12 steps are the only answer and frown on using the prescription drugs that have been shown to help people reduce their drinking.

People with alcohol problems also suffer from higher-than-normal rates of mental health issues, and research has shown that treating depression and anxiety with medication can reduce drinking. But AA is not equipped to address these issues—it is a support group whose leaders lack professional training—and some meetings are more accepting than others of the idea that members may need therapy and/or medication in addition to the group's help.

AA truisms have so infiltrated our culture that many people believe heavy drinkers cannot recover before they "hit bottom." Researchers I've talked with say that's akin to offering antidepressants only to those who have attempted suicide, or prescribing insulin only after a patient has lapsed into a diabetic coma. "You might as well tell a guy who weighs 250 pounds and has untreated hypertension and cholesterol of 300, 'Don't exercise, keep eating fast food, and we'll give you a triple bypass when you have a heart attack,'" Mark Willenbring, a psychiatrist in St. Paul and a former director of treatment and recovery research at the National Institute on Alcohol Abuse and Alcoholism, told me. He threw up his hands. "Absurd."

Part of the problem is our one-size-fits-all approach. Alcoholics Anonymous was originally intended for chronic, severe drinkers—those who may, indeed, be powerless over alcohol—but its program has since been applied much more broadly. Today, for instance, judges routinely require people to attend meetings after a DUI arrest; fully 12 percent of AA members are there by court order.

Whereas AA teaches that alcoholism is a progressive disease that follows an inevitable trajectory, data from a federally funded survey called the National Epidemiological Survey on Alcohol and Related Conditions show that nearly one-fifth of those who have had alcohol dependence go on to drink at low-risk levels with no symptoms of abuse. And a recent survey of nearly 140,000 adults by the Centers for Disease Control and Prevention found that 9 out of 10 heavy drinkers are not dependent on alcohol and, with the help of a medical professional's brief intervention, can change unhealthy habits.

We once thought about drinking problems in binary terms—you either had control or you didn't; you were an alcoholic or you weren't—but experts now describe a spectrum. An estimated 18 million Americans suffer from alcohol-use disorder, as the *DSM-5*, the latest edition of the American Psychiatric Association's diagnostic manual, calls it. (The new term replaces the older "alcohol abuse" and the much more dated "alcoholism," which has been out of favor with researchers for decades.) Only about 15 percent of those with alcohol-use disorder are at the severe end of the spectrum. The rest fall somewhere in the mild-to-moderate range, but they have been largely ignored by researchers and clinicians. Both groups—the hardcore abusers and the more moderate over-drinkers—need more-individualized treatment options.

The United States already spends about $35 billion a year on alcohol- and substance-abuse treatment, yet heavy drinking causes 88,000 deaths a year—including deaths from car accidents and diseases linked to alcohol. It also costs the country hundreds of billions of dollars in expenses related to health care, criminal justice, motor-vehicle crashes, and lost workplace productivity, according to the CDC. With the Affordable Care Act's expansion of coverage, it's time to ask some important questions: Which treatments should we be willing to pay for? Have they been proved effective? And for whom? Only those at the extreme end of the spectrum? Or also those in the vast, long-overlooked middle?

For a glimpse of how treatment works elsewhere, I traveled to Finland, a country that shares with the United States a history of prohibition (inspired by the American temperance movement, the Finns outlawed alcohol from 1919 to 1932) and a culture of heavy drinking.

Finland's treatment model is based in large part on the work of an American neuroscientist named John David Sinclair. I met with Sinclair in Helsinki in early July. He was battling late-stage prostate cancer, and his thick white hair was cropped short in preparation for chemotherapy. Sinclair has researched alcohol's effects on the brain since his days as an undergraduate at the University of Cincinnati, where he experimented with rats that had been given alcohol for an extended period. Sinclair expected that after several weeks without booze, the rats would lose their desire for it.

Instead, when he gave them alcohol again, they went on weeklong benders, drinking far more than they ever had before—more, he says, than any rat had ever been shown to drink.

Sinclair called this the alcohol-deprivation effect, and his laboratory results, which have since been confirmed by many other studies, suggested a fundamental flaw in abstinence-based treatment: going cold turkey only intensifies cravings. This discovery helped explain why relapses are common. Sinclair published his findings in a handful of journals and in the early 1970s moved to Finland, drawn by the chance to work in what he considered the best alcohol-research lab in the world, complete with special rats that had been bred to prefer alcohol to water. He spent the next decade researching alcohol and the brain.

Sinclair came to believe that people develop drinking problems through a chemical process: each time they drink, the endorphins released in the brain strengthen certain synapses. The stronger these synapses grow, the more likely the person is to think about, and eventually crave, alcohol—until almost anything can trigger a thirst for booze, and drinking becomes compulsive.

Sinclair theorized that if you could stop the endorphins from reaching their target, the brain's opiate receptors, you could gradually weaken the synapses, and the cravings would subside. To test this hypothesis, he administered opioid antagonists—drugs that block opiate receptors—to the specially bred alcohol-loving rats. He found that if the rats took the medication each time they were given alcohol, they gradually drank less and less. He published his findings in peer-reviewed journals beginning in the 1980s.

Subsequent studies found that an opioid antagonist called naltrexone was safe and effective for humans, and Sinclair began working with clinicians in Finland. He suggested prescribing naltrexone for patients to take an hour before drinking. As their cravings subsided, they could then learn to control their consumption. Numerous clinical trials have confirmed that the method is effective, and in 2001 Sinclair published a paper in the journal *Alcohol and Alcoholism* reporting a 78 percent success rate in helping patients reduce their drinking to about 10 drinks a week. Some stopped drinking entirely.

I visited one of three private treatment centers, called the Contral Clinics, that Sinclair cofounded in Finland. (There's an additional one in Spain.) In the past 18 years, more than 5,000 Finns have gone to the Contral Clinics for help with a drinking problem. Seventy-five percent of them have had success reducing their consumption to a safe level.

The Finns are famously private, so I had to go early in the morning, before any patients arrived, to meet Jukka Keski-Pukkila, the CEO. He poured coffee and showed me around the clinic, in downtown Helsinki. The most common course of treatment involves six months of cognitive-behavioral therapy, a goal-oriented form of therapy, with a clinical psychologist. Treatment typically also includes a physical exam, blood work, and a prescription for naltrexone or nalmefene, a newer opioid antagonist approved in more than two dozen countries. When I asked how much all of this cost, Keski-Pukkila looked uneasy. "Well," he told me, "it's 2,000 euros." That's about $2,500—a fraction of the cost of inpatient rehab in the United States, which routinely runs in the tens of thousands of dollars for a 28-day stay.

When I told Keski-Pukkila this, his eyes grew wide. "What are they doing for that money?" he asked. I listed some of the treatments offered at top-of-the-line rehab centers: equine therapy, art therapy, mindfulness mazes in the desert. "That doesn't sound scientific," he said, perplexed. I didn't mention that some bare-bones facilities charge as much as $40,000 a month and offer no treatment beyond AA sessions led by minimally qualified counselors.

As I researched this article, I wondered what it would be like to try naltrexone, which the U.S. Food and Drug Administration approved for alcohol-abuse treatment in 1994. I asked my doctor whether he would write me a prescription. Not surprisingly, he shook his head no. I don't have a drinking problem, and he said he couldn't offer medication for an "experiment." So that left the Internet, which was easy enough. I ordered some naltrexone online and received a foil-wrapped package of 10 pills about a week later. The cost was $39.

The first night, I took a pill at 6:30. An hour later, I sipped a glass of wine and felt almost nothing—no calming effect, none of the warm contentment that usually signals the end of my workday and the beginning of a relaxing evening. I finished the glass and poured a second. By the end of dinner, I looked up to see that I had barely touched it. I had never found wine so uninteresting. Was this a placebo effect? Possibly. But so it went. On the third night, at a restaurant where my husband and I split a bottle of wine, the waitress came to refill his glass twice; mine, not once. That had never happened before, except when I was pregnant. At the end of 10 days, I found I no longer looked forward to a glass of wine with dinner. (Interestingly, I also found myself feeling full much quicker than normal, and I lost two pounds. In Europe, an opioid antagonist is being tested on binge eaters.)

I was an *n* of one, of course. My experiment was driven by personal curiosity, not scientific inquiry. But it certainly felt as if I were unlearning something—the pleasure of that first glass? The desire for it? Both? I can't really say.

Patients on naltrexone have to be motivated to keep taking the pill. But Sari Castrén, a psychologist at the Contral Clinic I visited in Helsinki, told me that when patients come in for treatment, they're desperate to change the role alcohol has assumed in their lives. They've tried not drinking, and controlling their drinking, without success—their cravings are too strong. But with naltrexone or nalmefene, they're able to drink less, and the benefits soon become apparent: They sleep better. They have more energy and less guilt. They feel proud. They're able to read or watch movies or play with their children during the time they would have been drinking.

In therapy sessions, Castrén asks patients to weigh the pleasure of drinking against their enjoyment of these new activities, helping them to see the value of change. Still, the combination of naltrexone and therapy doesn't work for everyone. Some clients opt to take Antabuse, a medication that triggers nausea, dizziness, and other uncomfortable reactions when combined with drinking. And some patients are unable to learn how to drink without losing control. In those cases (about 10 percent of patients), Castrén recommends total abstinence from alcohol, but she leaves that choice to patients. "Sobriety is their decision, based on their own discovery," she told me.

Claudia Christian, an actress who lives in Los Angeles (she's best known for appearing in the 1990s science-fiction TV show *Babylon 5*), discovered naltrexone when she came across a flier for Vivitrol, an injectable form of the drug, at a detox center in California in 2009. She had tried Alcoholics Anonymous and traditional rehab without success. She researched the medication online, got a doctor to prescribe it, and began taking a dose about an hour before she planned to drink, as Sinclair recommends. She says the effect was like flipping a switch. For the first time in many years, she was able to have a single drink and then stop. She plans to keep taking naltrexone indefinitely, and has become an advocate for Sinclair's method: she set up a nonprofit organization for people seeking information about it and made a documentary called *One Little Pill.*

In the United States, doctors generally prescribe naltrexone for daily use and tell patients to avoid alcohol, instead of instructing them to take the drug anytime they plan to drink, as Sinclair would advise. There is disagreement among experts about which approach is better—Sinclair is adamant that American doctors are missing the drug's full potential—but both seem to work: naltrexone has been found to reduce drinking in more than a dozen clinical trials, including a large-scale one funded by the National Institute on Alcohol Abuse and Alcoholism that was published in *JAMA* in 2006. The results have been largely overlooked. Less than 1 percent of people treated for alcohol problems in the United States are prescribed naltrexone or any other drug shown to help control drinking.

To understand why, you have to first understand the history.

The American approach to treatment for drinking problems has roots in the country's long-standing love-hate relationship with booze. The first settlers arrived with a great thirst for whiskey and hard cider, and in the early days of the Republic, alcohol was one of the few beverages that was reliably safe from contamination. (It was also cheaper than coffee or tea.) The historian W. J. Rorabaugh has estimated that between the 1770s and 1830s the average American over age 15 consumed at least five gallons of pure alcohol a year—the rough equivalent of three shots of hard liquor a day.

Religious fervor, aided by the introduction of public water-filtration systems, helped galvanize the temperance movement, which culminated in 1920 with Prohibition. That experiment ended after 14 years, but the drinking culture it fostered—secrecy and frenzied bingeing—persists.

In 1934, just after Prohibition's repeal, a failed stockbroker named Bill Wilson staggered into a Manhattan hospital. Wilson was known to drink two quarts of whiskey a day, a habit he'd attempted to kick many times. He was given the hallucinogen belladonna, an experimental treatment for addictions, and from his hospital bed he called out to God to loosen alcohol's grip. He reported seeing a flash of light and feeling a serenity he had never before experienced. He quit booze for good. The next year, he cofounded Alcoholics Anonymous. He based its principles on the beliefs of the evangelical Oxford Group, which taught that people were sinners who, through confession and God's help, could right their paths.

AA filled a vacuum in the medical world, which at the time had few answers for heavy drinkers. In 1956 the American Medical Association named alcoholism a disease, but doctors continued to offer little beyond the standard treatment that had been around for decades: detoxification in state psychiatric wards or private sanatoriums. As Alcoholics Anonymous grew, hospitals began creating "alcoholism wards," where patients detoxed but were given no other medical treatment. Instead, AA members—who, as part of the 12 steps, pledge to help other alcoholics—appeared at bedsides and invited the newly sober to meetings.

A public relations specialist and early AA member named Marty Mann worked to disseminate the group's main tenet: that alcoholics had an illness that rendered them powerless over booze. Their drinking was a disease, in other words, not a moral failing. Paradoxically, the prescription for this medical condition was a set of spiritual steps that required accepting a higher power, taking a "fearless moral inventory," admitting "the exact nature of our wrongs," and asking God to remove all character defects.

Mann helped ensure that these ideas made their way to Hollywood. In 1945's *The Lost Weekend,* a struggling novelist tries to loosen his writer's block with booze, to devastating effect. In *Days of Wine and Roses,* released in 1962, Jack Lemmon slides into alcoholism along with his wife, played by Lee Remick. He finds help through AA, but she rejects the group and loses her family.

Mann also collaborated with a physiologist named E. M. Jellinek. Mann was eager to bolster the scientific claims behind AA, and Jellinek wanted to make a name for himself in the growing field of alcohol research. In 1946 Jellinek published the results of a survey mailed to 1,600 AA members. Only 158 were returned. Jellinek and Mann jettisoned 45 that had been improperly completed and another 15 filled out by women, whose responses were so unlike the men's that they risked complicating the results. From this small sample—98 men—Jellinek drew sweeping conclusions about the "phases of alcoholism," which included an unavoidable succession of binges that led to blackouts, "indefinable fears," and hitting bottom. Though the paper was filled with caveats about its lack of scientific rigor, it became AA gospel.

Jellinek, however, later tried to distance himself from this work, and from Alcoholics Anonymous. His ideas came to be illustrated by a chart showing how alcoholics progressed from occasionally drinking for relief, to sneaking drinks, to guilt, and so on until they hit bottom ("complete defeat admitted") and then recovered. If you could locate yourself even early in the downward trajectory on that curve, you could see where your drinking was headed. In 1952 Jellinek noted that the word "alcoholic" had been adopted to describe anyone who drank excessively. He warned that overuse of that word would undermine the disease concept. He later beseeched AA to stay out of the way of scientists trying to do objective research.

But AA supporters worked to make sure their approach remained central. Marty Mann joined prominent Americans including Susan Anthony, the grandniece of Susan B. Anthony; Jan Clayton, the mom from *Lassie;* and decorated military officers in testifying before Congress. John D. Rockefeller Jr., a lifelong teetotaler, was an early booster of the group.

In 1970 Senator Harold Hughes of Iowa, a member of AA, persuaded Congress to pass the Comprehensive Alcohol Abuse and Alcoholism Prevention, Treatment, and Rehabilitation Act. It called for the establishment of the National Institute on Alcohol Abuse and Alcoholism, and dedicated funding for the study and treatment of alcoholism. The NIAAA, in turn, funded Marty Mann's nonprofit advocacy group, the National Council on Alcoholism, to educate the public. The nonprofit became a mouthpiece for AA's beliefs, especially the importance of abstinence, and has at times worked to quash research that challenges those beliefs.

In 1976, for instance, the RAND Corporation released a study of more than 2,000 men who had been patients at 44 different NIAAA-funded treatment centers. The report noted that 18 months after treatment, 22 percent of the men were drinking moderately.

The authors concluded that it was possible for some alcohol-dependent men to return to controlled drinking. Researchers at the National Council on Alcoholism charged that the news would lead alcoholics to falsely believe they could drink safely. The NIAAA, which had funded the research, repudiated it. RAND repeated the study, this time looking over a four-year period. The results were similar.

After the Hughes Act was passed, insurers began to recognize alcoholism as a disease and pay for treatment. For-profit rehab facilities sprouted across the country, the beginnings of what would become a multibillion-dollar industry. (Hughes became a treatment entrepreneur himself, after retiring from the Senate.) If Betty Ford and Elizabeth Taylor could declare that they were alcoholics and seek help, so too could ordinary people who struggled with drinking. Today there are more than 13,000 rehab facilities in the United States, and 70 to 80 percent of them hew to the 12 steps, according to Anne M. Fletcher, the author of *Inside Rehab,* a 2013 book investigating the treatment industry.

The problem is that nothing about the 12-step approach draws on modern science: not the character building, not the tough love, not even the standard 28-day rehab stay.

Marvin D. Seppala, the chief medical officer at the Hazelden Betty Ford Foundation in Minnesota, one of the oldest inpatient rehab facilities in the country, described for me how 28 days became the norm: "In 1949 the founders found that it took about a week to get detoxed, another week to come around so [the patients] knew what they were up to, and after a couple of weeks they were doing well, and stable. That's how it turned out to be 28 days. There's no magic in it."

Tom McLellan, a psychology professor at the University of Pennsylvania School of Medicine who has served as a deputy U.S. drug czar and is an adviser to the World Health Organization, says that while AA and other programs that focus on behavioral change have value, they don't address what we now know about the biology of drinking.

Alcohol acts on many parts of the brain, making it in some ways more complex than drugs like cocaine and heroin, which target just one area of the brain. Among other effects, alcohol increases the amount of GABA (gamma-aminobutyric acid), a chemical that slows down activity in the nervous system, and decreases the flow of glutamate, which activates the nervous system. (This is why drinking can make you relax, shed inhibitions, and forget your worries.) Alcohol also prompts the brain to release dopamine, a chemical associated with pleasure.

Over time, though, the brain of a heavy drinker adjusts to the steady flow of alcohol by producing less GABA and more glutamate, resulting in anxiety and irritability. Dopamine production also slows, and the person gets less pleasure out of everyday things. Combined, these changes gradually bring about a crucial shift: instead of drinking to feel good, the person ends up drinking to avoid feeling bad. Alcohol also damages the prefrontal cortex, which is responsible for judging risks and regulating behavior—one reason some people keep drinking even as they realize that the habit is destroying their lives. The good news is that the damage can be undone if they're able to get their consumption under control.

Studies of twins and adopted children suggest that about half of a person's vulnerability to alcohol-use disorder is hereditary, and that anxiety, depression, and environment—all considered "outside issues" by many in Alcoholics Anonymous and the rehab industry—also play a role. Still, science can't yet fully explain why some heavy drinkers become physiologically dependent on alcohol and others don't, or why some recover while others founder. We don't know how much drinking it takes to cause major changes in the brain, or whether the brains of alcohol-dependent people are in some ways different from "normal" brains to begin with. What we do know, McLellan says, is that "the brains of the alcohol-addicted aren't like those of the non-alcohol-dependent."

Bill Wilson, AA's founding father, was right when he insisted, 80 years ago, that alcohol dependence is an illness, not a moral failing. Why, then, do we so rarely treat it medically? It's a question I've heard many times from researchers and clinicians. "Alcohol- and substance-use disorders are the realm of medicine," McLellan says. "This is not the realm of priests."

When the Hazelden treatment center opened in 1949, it espoused five goals for its patients: behave responsibly, attend lectures on the 12 steps, make your bed, stay sober, and talk with other patients. Even today, Hazelden's website states:

> People addicted to alcohol can be secretive, self-centered, and filled with resentment. In response, Hazelden's founders insisted that patients attend to the details of daily life, tell their stories, and listen to each other . . . This led to a heartening discovery, one that's become a cornerstone of the Minnesota Model: Alcoholics and addicts can help each other.

That may be heartening, but it's not science. As the rehab industry began expanding in the 1970s, its profit motives dovetailed nicely with AA's view that counseling could be delivered by people who had themselves struggled with addiction, rather than by highly trained (and highly paid) doctors and mental health professionals. No other area of medicine or counseling makes such allowances.

There is no mandatory national certification exam for addiction counselors. The 2012 Columbia University report on addiction medicine found that only six states required alcohol- and substance-abuse counselors to have at least a bachelor's degree and that only one state, Vermont, required a master's degree. Fourteen states had no license requirements whatsoever—not even a GED or an introductory training course was necessary—and yet counselors are often called on by the judicial system and medical boards to give expert opinions on their clients' prospects for recovery.

Mark Willenbring, the St. Paul psychiatrist, winced when I mentioned this. "What's wrong," he asked me rhetorically, "with people with no qualifications or talents—other than being recovering alcoholics—being licensed as professionals with decision-making authority over whether you are imprisoned or lose your medical license?

"The history—and current state—is really, really dismal," Willenbring said.

Perhaps even worse is the pace of research on drugs to treat alcohol-use disorder. The FDA has approved just three: Antabuse, the drug that induces nausea and dizziness when taken with alcohol; acamprosate, which has been shown to be helpful in quelling cravings; and naltrexone. (There is also Vivitrol, the injectable form of naltrexone.)

Reid K. Hester, a psychologist and the director of research at Behavior Therapy Associates, an organization of psychologists in

Albuquerque, says there has long been resistance in the United States to the idea that alcohol-use disorder can be treated with drugs. For a brief period, DuPont, which held the patent for naltrexone when the FDA approved it for alcohol-abuse treatment in 1994, paid Hester to speak about the drug at medical conferences. "The reaction was always 'How can you be giving alcoholics drugs?'" he recalls.

Hester says this attitude dates to the 1950s and '60s, when psychiatrists regularly prescribed heavy drinkers Valium and other sedatives with great potential for abuse. Many patients wound up dependent on both booze and benzodiazepines. "They'd look at me like I was promoting *Valley of the Dolls 2.0*," Hester says.

There has been some progress: the Hazelden center began prescribing naltrexone and acamprosate to patients in 2003. But this makes Hazelden a pioneer among rehab centers. "Everyone has a bias," Marvin Seppala, the chief medical officer, told me. "I honestly thought AA was the only way anyone could ever get sober, but I learned that I was wrong."

Stephanie O'Malley, a clinical researcher in psychiatry at Yale who has studied the use of naltrexone and other drugs for alcohol-use disorder for more than two decades, says naltrexone's limited use is "baffling."

"There was never any campaign for this medication that said, 'Ask your doctor,'" she says. "There was never any attempt to reach consumers." Few doctors accepted that it was possible to treat alcohol-use disorder with a pill. And now that naltrexone is available in an inexpensive generic form, pharmaceutical companies have little incentive to promote it.

In one recent study, O'Malley found naltrexone to be effective in limiting consumption among college-age drinkers. The drug helped subjects keep from going over the legal threshold for intoxication, a blood alcohol content of 0.08 percent. Naltrexone is not a silver bullet, though. We don't yet know for whom it works best. Other drugs could help fill in the gaps. O'Malley and other researchers have found, for example, that the smoking-cessation medication varenicline has shown promise in reducing drinking. So, too, have topiramate, a seizure medication, and baclofen, a muscle relaxant. "Some of these drugs should be considered in primary-care offices," O'Malley says. "And they're just not."

*

In late August I visited Alltyr, a clinic that Willenbring founded in St. Paul. It was here that J.G. finally found help.

After his stays in rehab, J.G. kept searching for alternatives to 12-step programs. He read about baclofen and how it might ease both anxiety and cravings for alcohol, but his doctor wouldn't prescribe it. In his desperation, J.G. turned to a Chicago psychiatrist who wrote him a prescription for baclofen without ever meeting him in person and eventually had his license suspended. Then, in late 2013, J.G.'s wife came across Alltyr's website and discovered, 20 minutes from his law office, a nationally known expert in treating alcohol- and substance-use disorders.

J.G. now sees Willenbring once every 12 weeks. During those sessions, Willenbring checks on J.G.'s sleep patterns and refills his prescription for baclofen (Willenbring was familiar with the studies on baclofen and alcohol, and agreed it was a viable treatment option), and occasionally prescribes Valium for his anxiety. J.G. doesn't drink at all these days, though he doesn't rule out the possibility of having a beer every now and then in the future.

I also talked with another Alltyr patient, Jean, a Minnesota floral designer in her late 50s who at the time was seeing Willenbring three or four times a month but has since cut back to once every few months. "I actually look forward to going," she told me. At age 50, Jean (who asked to be identified by her middle name) went through a difficult move and a career change, and she began soothing her regrets with a bottle of red wine a day. When Jean confessed her habit to her doctor last year, she was referred to an addiction counselor. At the end of the first session, the counselor gave Jean a diagnosis: "You're a drunk," he told her, and suggested she attend AA.

The whole idea made Jean uncomfortable. How did people get better by recounting the worst moments of their lives to strangers? Still, she went. Each member's story seemed worse than the last: One man had crashed his car into a telephone pole. Another described his abusive blackouts. One woman carried the guilt of having a child with fetal alcohol syndrome. "Everybody talked about their 'alcoholic brain' and how their 'disease' made them act," Jean told me. She couldn't relate. She didn't believe her affection for pinot noir was a disease, and she bristled at the lines people read from the Big Book: "We thought we could find a softer, easier way," they recited. "But we could not."

Surely, Jean thought, modern medicine had to offer a more current form of help.

Then she found Willenbring. During her sessions with him, she talks about troubling memories that she believes helped ratchet up her drinking. She has occasionally had a drink; Willenbring calls this "research," not "a relapse." "There's no belittling, no labels, no judgment, no book to carry around, no taking away your 'medal,'" Jean says, a reference to the chips that AA members earn when they reach certain sobriety milestones.

In his treatment, Willenbring uses a mix of behavioral approaches and medication. Moderate drinking is not a possibility for every patient, and he weighs many factors when deciding whether to recommend lifelong abstinence. He is unlikely to consider moderation as a goal for patients with severe alcohol-use disorder. (According to the *DSM-5,* patients in the severe range have six or more symptoms of the disorder, such as frequently drinking more than intended, increased tolerance, unsuccessful attempts to cut back, cravings, missing obligations due to drinking, and continuing to drink despite negative personal or social consequences.) Nor is he apt to suggest moderation for patients who have mood, anxiety, or personality disorders; chronic pain; or a lack of social support. "We can provide treatment based on the stage where patients are," Willenbring said. It's a radical departure from issuing the same prescription to everyone.

The difficulty of determining which patients are good candidates for moderation is an important cautionary note. But promoting abstinence as the only valid goal of treatment likely deters people with mild or moderate alcohol-use disorder from seeking help. The prospect of never taking another sip is daunting, to say the least. It comes with social costs and may even be worse for one's health than moderate drinking: research has found that having a drink or two a day could reduce the risk of heart disease, dementia, and diabetes.

To many, though, the idea of nonabstinent recovery is anathema.

No one knows that better than Mark and Linda Sobell, who are both psychologists. In the 1970s the couple conducted a study with a group of 20 patients in Southern California who had been diagnosed with alcohol dependence. Over the course of 17 sessions, they taught the patients how to identify their triggers, how to refuse drinks, and other strategies to help them drink safely. In a follow-up study two years later, the patients had fewer days of heavy drinking, and more days of no drinking, than did a group of 20 alcohol-dependent patients who were told to abstain from drinking entirely. (Both groups were given a standard hospital treatment, which included group therapy, AA meetings, and medications.) The Sobells published their findings in peer-reviewed journals.

In 1980 the University of Toronto recruited the couple to conduct research at its prestigious Addiction Research Foundation. "We didn't set out to challenge tradition," Mark Sobell told me. "We just set out to do good research." Not everyone saw it that way. In 1982 abstinence-only proponents attacked the Sobells in the journal *Science;* one of the writers, a UCLA psychologist named Irving Maltzman, later accused them of faking their results. The *Science* article received widespread attention, including a story in the *New York Times* and a segment on *60 Minutes.*

Over the next several years, four panels of investigators in the United States and Canada cleared the couple of the accusations. Their studies were accurate. But the exonerations had scant impact, Mark Sobell said: "Maybe a paragraph on page fourteen" of the newspaper.

The late G. Alan Marlatt, a respected addiction researcher at the University of Washington, commented on the controversy in a 1983 article in *American Psychologist.* "Despite the fact that the basic tenets of [AA's] disease model have yet to be verified scientifically," Marlatt wrote, "advocates of the disease model continue to insist that alcoholism is a unitary disorder, a progressive disease that can only be arrested temporarily by total abstention."

What's stunning, 32 years later, is how little has changed.

The Sobells returned to the United States in the mid-1990s to teach and conduct research at Nova Southeastern University, in Fort Lauderdale, Florida. They also run a clinic. Like Willenbring in Minnesota, they are among a small number of researchers and clinicians, mostly in large cities, who help some patients learn to drink in moderation.

"We cling to this one-size-fits-all theory even when a person has a small problem," Mark Sobell told me. "The idea is 'Well, this may be the person you are now, but this is where this is going, and there's only one way to fix it.'" Sobell paused. "But we have fifty years of research saying that, chances are, that's not the way it's going. We can change the course."

During my visit to Finland, I interviewed P., a former Contral Clinic patient who asked me to use only his last initial in order to protect his privacy. He told me that for years he had drunk to excess, sometimes having as many as 20 drinks at a time. A 38-year-old doctor and university researcher, he describes himself as mild-mannered while sober. When drunk, though, "it was as if some primitive human took over."

His wife found a Contral Clinic online, and P. agreed to go. From his first dose of naltrexone, he felt different—in control of his consumption for the first time. P. plans to use naltrexone for the rest of his life. He drinks two, maybe three, times a month. By American standards, these episodes count as binges, since he sometimes downs more than 5 drinks in one sitting. But that's a steep decline from the 80 drinks a month he consumed before he began the treatment—and in Finnish eyes, it's a success.

Sari Castrén, the psychologist I met at Contral, says such trajectories are the rule among her patients. "Helping them find this path is so rewarding," she says. "This is a softer way to look at addiction. It doesn't have to be so black-and-white."

J.G. agrees. He feels much more confident and stable, he says, than he did when he was drinking. He has successfully drunk in moderation on occasion, without any loss of control or desire to consume more the next day. But for the time being, he's content not drinking. "It feels like a big risk," he says. And he has more at stake now—his daughter was born in June 2013, about six months before he found Willenbring.

Could the Affordable Care Act's expansion of coverage prompt us to rethink how we treat alcohol-use disorder? That remains to be seen. The Department of Health and Human Services, the primary administrator of the act, is currently evaluating treatments. But the legislation does not specify a process for deciding which methods should be approved, so states and insurance companies are setting their own rules. How they'll make those decisions is a matter of ongoing discussion.

Still, many leaders in the field are hopeful—including Tom McLellan, the University of Pennsylvania psychologist. His optimism is particularly poignant: in 2008 he lost a son to a drug overdose. "If I didn't know what to do for my kid, when I know this stuff and am surrounded by experts, how the hell is a schoolteacher or a construction worker going to know?" he asks. Americans need to demand better, McLellan says, just as they did with breast cancer, HIV, and mental illness. "This is going to be a mandated benefit, and insurance companies are going to want to pay for things that work," he says. "Change is within reach."

ANTONIA JUHASZ

Thirty Million Gallons Under the Sea

FROM *Harper's Magazine*

ONE MORNING IN MARCH of last year, I set out from Gulfport, Mississippi, on a three-week mission aboard the U.S. Navy research vessel *Atlantis*. The 274-foot ship, painted a crisp white and blue, stood tall in the bright sunlight. On its decks were winches, cranes, seafloor-mapping sonar, a machine shop, and five laboratories. Stowed in an alcove astern was *Alvin,* the federal government's only manned research submarine. "Research vessel *Atlantis* outbound," A. D. Colburn, the ship's captain, reported into the ship radio.

The water was calm and the bridge crew quiet as they steered us into open water. For the next 14 hours, we would sail toward the site of BP's Macondo well, where in April 2010 a blowout caused the largest offshore-drilling oil spill in history. Once there, *Atlantis*'s crew would launch *Alvin* and guide it to the bottom of the ocean, reaching depths as great as 7,200 feet below the surface. Over the next 22 days they would send the submersible down 17 times, to gather animal, plant, water, and sediment samples. Their goal was to determine how BP's spill had affected the ocean's ecosystem from the seabed up. I would get the chance to join them in the submarine as they went closer to the Macondo wellhead than anyone had gone since the blowout.

Data gathered by the *Atlantis* would likely be used in the federal legal proceedings against BP, which began in December 2010. A few months after our mission, U.S. district judge Carl Barbier found the company guilty of gross negligence and willful misconduct. In January 2015 he ruled that the amount of oil the company was responsible for releasing into the Gulf totaled some 134 million gallons, a decision both sides have appealed. By the time this article went to press, Barbier had yet to make his third and final ruling, which will determine how much BP owes in penalties under the Clean Water Act. (If his judgment about the size of the spill is not overturned, the company will face a $13.7 billion fine.) Meanwhile, the Environmental Protection Agency, the National Oceanic and Atmospheric Administration, the Department of Agriculture, and the Department of the Interior are concluding an ecological-damages assessment to determine how much BP must pay to restore the Gulf Coast. The trial and the assessment are likely to result in the largest penalty ever leveled against an oil company.*

Dr. Samantha Joye, a biogeochemist at the University of Georgia and the lead scientist on the mission, estimated that 30 million gallons of oil from the BP spill remain in the Gulf—the equivalent of nearly three *Exxon Valdez* spills—and that about half of this amount has settled on the ocean floor, where its ecological effects could be devastating. A BP spokesman told me that most of the spilled oil had been removed, consumed, or degraded, and that there had been "very limited impact from the oil spill on the seafloor." But Joye's research indicates that any damage the oil has done to creatures inhabiting the deep-sea waters—from tubeworms to sperm whales—threatens the ecosystem, harming organisms that rely on those species. Among those at risk are phytoplankton, the sea vegetation that produces about half of the planet's oxygen. "If you short-circuit the bottom, you threaten the entire cycle," Joye told me. "Without a healthy ocean, we'll all be dead."

The Gulf of Mexico is one of the world's most ecologically diverse bodies of water, home to more than 15,000 marine species. Its coastal areas contain half of the wetlands in the United States—some 5 million acres—which provide habitat for a variety of birds and fish. Mangrove forests line the Gulf shore, and its shallow wa-

* Some of the funding for the *Atlantis*'s mission came from a $500 million initiative to study the effects of the spill. The program was endowed by BP in 2010, under pressure from the White House and Congress.

ters are filled with seagrass. Before the oil spill, more than 60 percent of all oysters harvested in U.S. waters were caught in the Gulf. Today that share has dropped to about 40 percent. Many threatened and endangered species also live in the Gulf: sea turtles, Florida manatees, whooping cranes, and bald eagles. Dolphins are a frequent sight. Giant squid, jellyfish, and octopuses swimming through the Gulf's waters pass some of America's lushest and most imperiled coral reefs.

As *Atlantis* carried us farther from shore, the fishing boats dropped away and we gradually entered clear, open ocean. Mississippi and Florida are the only Gulf states that do not allow offshore drilling in their coastal waters, which extend several miles out from shore. But as soon as we entered federal waters, oil rigs and drilling platforms began to appear. We sailed through the rigs for 12 hours and, near 11:00 at night, arrived at our destination, Mississippi Canyon Block 252 (MC252). This was the roughly 5,760-acre lease area that contained the Macondo oil well.

Much of the western and central Gulf of Mexico has been parceled into a patchwork of oil and natural-gas lease blocks. A web of underwater pipelines carries the fuel to shore. MC252 sits in an industrial corridor that is occupied by the world's largest oil companies—including ExxonMobil, Chevron, Shell, Eni, Noble Energy, Hess, and BP. Some 17 percent of U.S. oil comes from the Gulf, nearly 80 percent of that from depths of 1,000 feet or more below the ocean's surface. In 2008 BP leased MC252 from the Department of the Interior, paying $34 million for 10 years. The company then leased the Deepwater Horizon, a semisubmersible oil rig, from Transocean, which also ran the rig's daily operations. On February 15, 2010, the crew of the Deepwater Horizon started drilling at the Macondo site, some 5,000 feet below the ocean's surface. Eventually they dug through more than two miles of rock and sediment, to a depth of 18,360 feet.

The Gulf of Mexico's high concentrations of methane, along with other natural features, make it an especially dangerous place to drill. Natural gas, which is lighter than oil, can get into oil pipes and overwhelm a well's pressure-control systems, leading to a blowout. At 9:45 on the night of April 20, 2010, as the crew worked to prepare the well, methane escaped from the Macondo and shot up the steel pipe that connected it to the Deepwater Horizon. Inside the rig, sparks from machinery ignited the gas, setting off a series of explosions. The two men who were working near the pipe, in the rig's mudroom, were quickly incinerated. They were the first of 11 crew members to die. At 10:21 a.m. on April 22, the Deepwater Horizon collapsed into the ocean. Its wreckage remains strewn around the well to this day.

When the Macondo well blew, none of the oil companies operating in the deep waters of the Gulf were prepared, even though the largest among them—ExxonMobil, Chevron, Shell, and BP—had claimed to the Department of the Interior that they could handle a far-worse deepwater blowout. So BP applied methods designed for smaller, shallow-water spills. In May, employing what were known as the "top kill" and "junk shot" approaches, BP dumped drilling mud, golf balls, and tire rubber onto the well. Nothing worked. In the 87 days it took to secure the well with a temporary cap, more than a hundred million gallons of oil and half a million tons of natural gas—most of which was methane—escaped into the Gulf. Eventually, BP performed the one operation that, however risky and time-consuming, it knew how to do: it drilled another well. One hundred and fifty-two days after the blowout, BP's relief well intersected the Macondo borehole, allowing the company to pump in mud and cement. This remains the only proven method for permanently sealing a blown-out well in deep water.

By May 2011 BP's oil had sickened or killed more than 100,000 Gulf animals: 28,500 sea turtles, 82,000 birds, and more than 26,000 marine mammals, including several sperm whales. Too small or too numerous to count were the vast numbers of dead fish, crustaceans, insects, and plants that washed up onshore. Most of the other organisms initially killed by the spill died at sea and were never seen.

The creatures that inhabited the coldest and darkest realms of the Gulf were not spared, either. Until the 19th century, when a pioneering British naval expedition was able to collect living samples from the seafloor, there was no evidence that animals could survive in the ocean's deepest waters. In 1977, more than a century later, researchers used *Alvin* to explore the Galápagos Rift, in the Pacific Ocean, at depths never before visited. To their surprise, they observed a broad diversity of life, including many previously unknown species. A 1984 *Alvin* dive revealed abundant populations thriving in the deepest parts of the Gulf, as well. In

December 2010 Joye and her crew surveyed the sea life near the Macondo wellhead. The view from the submarine revealed a barren landscape. The spill had chased out or killed anything that had been living down there. "It's so strange to see nothing along the seafloor, particularly at this depth and in this area," Joye wrote in her 2010 *Alvin* dive report. "I saw nothing on the bottom that was living," she told me.

When we arrived at MC252, *Atlantis*'s 24-person science party, most of them women in their 20s, immediately got to work. One group began deploying equipment from the decks, including sampling tools so large they had to be lowered into the water by crane. A device called a MOCNESS—a Multiple Opening/Closing Net and Environmental Sampling System—floated alongside the ship like a green sea monster, gathering tiny creatures to analyze how oil and gas were being taken up into the food web. Joye and her team disappeared belowdecks to prepare for the next day's dive.

Life aboard *Atlantis* revolved around *Alvin*. A 12-person team of pilots, engineers, and technicians was constantly at work on the sub. *Alvin* was commissioned by the Woods Hole Oceanographic Institution (which continues to operate it, on behalf of the Navy) and built by General Mills in 1964, six years after the establishment of the U.S. space program. In 1966 *Alvin* located a lost hydrogen bomb in the Mediterranean Sea; in 1986 the sub surveyed the wreckage of the *Titanic*. For many of the scientists, a key payoff for their long hours and grueling working conditions (they rarely slept during their three weeks at sea) was the chance that they would get to take part in a dive.

Since Joye's 2010 mission, *Alvin* had undergone a three-year overhaul that replaced 75 percent of its components and cost $41 million. Despite the work, the sub proved finicky. On a test dive a few weeks before we set sail, the oxygen scrubber, which removes carbon dioxide from the air, had failed, requiring *Alvin*'s three passengers to put on oxygen masks during their rapid return to the surface. Technicians later fashioned a simple umbrellalike device to temporarily fix the problem—just one of many contrivances that kept the sub running. (In a workroom one day I found a box filled with condoms. An *Alvin* technician explained that they were especially useful for protecting electrical wiring from exposure to seawater.)

Alvin is 23 feet long, round and squat with a bright white shell, a red hatch, and long metal arms reaching from its sides. The sub was named for Allyn Vine, a pioneering submersibles engineer, but members of its crew, following nautical tradition, refer to *Alvin* as "she," or sometimes "the Ball." They never call *Alvin* "it." The language they use is that of a very demanding and complicated personal relationship. At a media event the day before we left port, *Alvin* had appeared on deck, slowly rolling out on yellow tracks. I caught myself and several other reporters turning our recorders and microphones toward her, as if we expected her to speak.

By five o'clock on the morning of April 1, the day I was scheduled to dive, the *Alvin* crew was already up and preparing the sub. My dive would be the second of the trip, and the first to approach the Macondo wellhead.

After it rolled out to the stern's edge at seven o'clock, *Alvin* was tethered to a mechanical winch with a thick white rope that would lower it into the water. During my time on the *Atlantis* I had learned how to maneuver my way into an oversize red rubber survival suit (the "Gumby suit") in case we had to abandon ship. I'd learned, too, how to use the emergency-breathing device, which looked like a cross between a 1940s gas mask and a canister vacuum cleaner, in the event that the sub's scrubber failed again. I knew not to bring shoes on the dive, and to wear thick socks and warm clothing, because we'd be sitting still for eight hours. It gets very cold at the bottom of the ocean.

Bruce Strickrott, the *Atlantis* expedition leader, became an *Alvin* pilot in 1997, after retiring from the Navy. He compared diving in the sub to space travel, and *Alvin*'s silver and black interior does indeed resemble a spacecraft. The top half is packed with computers, monitors, electrical cables, and dashboards covered with red and black buttons and silver switches. Most of the sub's volume is given over to thrusters, banks of lead-acid batteries, and bundles of electrical wiring. Its three passengers share a sphere seven feet in diameter. Inside the sphere are five viewing portholes—three at the sub's nose, in front of the pilot, and one on each side—and the walls are draped in black insulated fabric to keep out the cold. As Joye, Bob Waters—the pilot for my dive—and I entered the sub in socks, sweats, and wool hats, I felt as though we were settling into a cozy camping tent. A more sinister feeling crept in, however, when *Alvin*'s ladder was drawn up and the hatch closed securely behind us.

After Strickrott gave us permission to depart, we gently descended underwater. Through my tiny circular window I saw the sunlight and clouds give way to bubbling blue liquid.

"Damn it!" Waters grumbled, after we'd dropped just a few feet below the surface. Out the front window, we watched as a pair of milk crates, which had been secured to the front of the sub to carry the tubes used to collect samples, sank into the deep. A pair of crew members in swim trunks, riding atop the sub in case of just this sort of situation, dove after the crates. They saved one, but the other got away.

Once the salvaged crate was reattached, Strickrott's soothing voice came back over the radio. "Let's do this again," he said, and we descended once more.

Joye and I are both small, and we sat with our legs stretched out on the cushioned floor. Waters crouched on a low stool that faced the computers he used to pilot *Alvin.* Once we reached the bottom, we were to identify any visual changes from the December 2010 dive and to collect water and sediment samples for further study. The scientists from Joye's research consortium wanted to know how much oil was on the ocean floor, whether microbes were consuming the oil that remained, and how the sea life was faring in the depths.

Our descent took two hours, at a pace that felt motionless. We could see no farther than a few feet from our windows, and there was little change in the scenery. Waters reported regularly to the *Atlantis* via an underwater telephone, and the digital display on the fathometer marked off our continuing dive, but the only visible sign of our depth was the color of the water—powder-blue became turquoise and then navy before fading into total darkness.

At 1,223 feet, the gloom was suddenly illuminated by zooplankton, which appeared as sinuous black lines with glowing tops. Joye explained that these tiny animals hosted luminescent bacteria that lit up when the zooplankton were surprised or alarmed. When we reached 3,609 feet, two-thirds of the way down, I asked Waters if he was excited.

"About what?" he responded. This was his 120th dive, give or take. He'd started working with *Alvin* in 1995, after building a laser-tag system for a Defense Department training program. Much of the rest of the *Alvin* crew was also composed of engineers, because, as Waters told me, "You've got to be able to fix it, not just fly it."

At 5,272 feet, we hit bottom. *Alvin*'s LEDs came on. If there was oil down here, we couldn't see it. Outside was an endless gray underwater desert: barren, flat, and stark. Through our tinted windows it looked like a vast moonscape.

Our dive brought us within two nautical miles of the wellhead. Any nearer and we would have risked getting caught in the wreckage of the Deepwater Horizon. We traveled about a mile and a half in five hours, tracing half a circle around the site, and stopped periodically to collect samples. By manipulating *Alvin*'s robotic arms and fingers, Waters was able to remove each sample tube from the crate, push it into the sediment, and delicately return it to its place.

On the previous day's dive, Joye and Joseph Montoya, a biological oceanographer at Georgia Tech, had been disturbed by their observation of dead and damaged coral—healthy coral provides habitat for thousands of species; dead coral is home to nothing. Today, however, there was a bit of good news: we saw a handful of sea cucumbers, small white fish, red crabs, blue eels, and pink shrimp. Etched along the seafloor we noted little pencil-shaped lines, evidence of organisms called infauna, which burrow, wormlike, into the sediment. As we passed, sea creatures struck attack poses: eels hung vertically in the water; crabs extended their claws and hind legs in our direction. Joye exclaimed at the sight of a vampire squid, a rare cephalopod, which showed us its bright red body and small webbed tentacles as it sped past. Later we saw a giant isopod, which Joye described as a "swimming cockroach." These gave her hope. But she told me that before the oil spill, we would have expected to see many more of these and other creatures: fish, urchins, sea fans, and perhaps even whales and sharks.

After a few hours we stopped to eat lunch. Outside the sub, it was 39 degrees Fahrenheit. The cold seeped in, causing condensation along the walls that soaked anything pressed against them.

Shortly after we resumed our journey, the flat seabed topography was broken by a set of mounds running in straight, parallel lines. They were too symmetrical not to be human-made. "Someone's been messing around down here," said Joye. We followed the lines for a while, then returned to our sampling.

At three in the afternoon, Waters announced the completion of our mission to the *Atlantis,* which gave us clearance to ascend. The sub had no toilet, so Joye, who was in need of one, was forced to use the dreaded pee bottle. I held up a blanket, Waters turned up the music—Adele—and Joye regaled us with the story of "the first guy to poop in the sub." When we reached the surface, we spent 20 minutes bouncing in the waves until the *Atlantis*'s swimmers were able to hook us onto the winch. Once we were back on deck, Montoya and another researcher dumped two giant buckets of seawater over my head, initiating me into the exclusive club of *Alvin* divers.

Over the next 10 days, I spent dozens of hours in the ship's labs, watching the researchers sort and analyze specimens. The sediment samples we'd gathered were, it turned out, virtually identical to the ones collected in 2010. The layer of oil residue deposited four years earlier was still there. "It looks the same no matter where you are," Joye said. "And it hasn't changed."

Today a coating of degraded oil, as much as two inches thick, extends across nearly 3,000 square miles of ocean floor. It is expected to remain there forever. In the *Atlantis*'s computer lab, Andreas Teske, a microbial ecologist at the University of North Carolina at Chapel Hill, told me, "When another expedition comes here in a hundred or a thousand years, they will say, 'Ah, OK. Here is the 2010 oil spill.'"

There are many reasons that oil remains on the seafloor. The cold, dark bottom of the ocean is a naturally preservative environment. In addition, Joye found that microbes that consumed some of the oil and methane in the first few months after the spill have largely stopped eating. What they've left behind—the parts they have not yet broken down—are among the most toxic components of the oil, including polycyclic aromatic hydrocarbons, which are known human carcinogens. The microbes have also been inhibited by Corexit, a toxic chemical dispersant that was supposed to keep the oil from drifting ashore. Nearly two million gallons of the dispersant were used in the aftermath of the spill. Joye's research now suggests, however, that Corexit was not only environmentally harmful; it was also counterproductive.

What does it mean that a blanket of oil remnants will cover thousands of miles of ocean floor for the foreseeable future? "We don't know exactly," Joye told me. Nothing close to the size and duration of this disaster has ever been studied, and it will take years to fully understand the spill's effects. But the data collected so far is alarming. Last June I talked to Dr. Paul Montagna, a marine ecologist at Texas A&M University who studies benthic organisms. He had found significant declines in a range of species that live on the Gulf seafloor. Within a 9-square-mile area around the Macondo wellhead, he measured a 50 percent loss in the biodiversity of tiny invertebrates called meiofauna and slightly larger species called macrofauna. These species are a critical food source for larger organisms. Meiofauna and macrofauna have suffered losses as far as 10 miles from the well. A die-off at any link in the food web threatens the species that depend on it, but it can also affect those farther down. Phytoplankton, for instance, rely on seafloor macrofauna such as tubeworms to help decompose organic matter and release nutrients back into the water.

The increase in sea life that we had observed on *Alvin* could signal the start of an ecological recovery, Montagna said. But those returning species were now also being exposed to the oil on the seafloor, which they would pass along to the creatures that ate them. Joe Montoya's research on phytoplankton has uncovered clear evidence that oil and gas carbon are moving through the food web. Ultimately, these contaminants, in potentially harmful concentrations, could reach "things like big fish that people are commercially interested in," Montoya told me.

One study has already demonstrated that the spill was followed by immediate declines in the larval production of tuna, blue marlin, mahi-mahi, and sailfish. Macondo oil has also been linked to life-threatening heart defects in embryonic and juvenile bluefin and yellowfin tuna, as well as in amberjack. Perhaps even more troubling has been the effect on dolphins, which are predators at the top of the food chain and therefore indicators of the ecosystem's degradation. In 2011 dolphins were stillborn or died in infancy at rates six times the historical average. Last year, the number of dolphins found dead on the Louisiana coast was four times higher than the annual average before the spill.

The effects of BP's disaster have now spread far from the Gulf. Traces of oil and Corexit, for instance, have been found in Minnesota, Iowa, and Illinois, in the eggs of white pelicans that were in the Gulf at the time of the spill. Nor are humans immune to the damage. In January 2013 BP agreed to a medical-benefits settlement that provides 21 years of health monitoring and potential monetary compensation—up to $60,000 per person—to Gulf Coast residents and cleanup workers who can demonstrate spill-related respiratory, gastrointestinal, eye, skin, and neurophysiological conditions. (Researchers have also found that crude oil contamination can lead to cancer, birth defects, and developmental and neurological disorders such as dementia, though none of these are covered by the settlement.) Of the more than 200,000 people who were potentially eligible for remuneration, however, only 12,144 had filed claims by the end of 2014, a spokesperson for the court-appointed medical-benefits-claims administrator told me. Of those, a mere 1,304 have been approved for payment.*

In June of last year a tar mat composed of degraded BP oil that weighed more than a thousand pounds was found on a beach near Fort Pickens, Florida. BP's oil also remains lodged in shoreline marshes, where it is killing plants and intensifying coastal erosion. In March another tar mat, this one weighing 29,000 pounds, was found buried in the marshes of East Grand Terre Island, Louisiana. The beaches of Bay Jimmy and Bay Batiste, also in Louisiana, are still so heavily oiled that they remain closed to shrimping, crabbing, fin fishing, and oystering. "Our catch is down by a hundred percent," Byron Encalade, a Louisiana oysterman, told me in March. And although the Food and Drug Administration declared in fall 2010 that many Gulf fish products were safe for human consumption, one 2011 study found that people who ate a diet heavy in Gulf seafood (or who were medically vulnerable) could be at risk of developmental disorders and cancer. "When people say, 'Oh, the oil spill is over,'" Joye told me, "they're not realizing that the full impacts are on a very long timescale, of decades or more."

Though there has been no drilling in MC252 since the Macondo was sealed, this won't be the case for long. During my *Alvin* dive, Waters had steered the sub to follow the suspicious mounds we'd seen running along the seafloor. They looked like freshly laid receiver cables, which are used in seismic surveys to map underground reserves of oil and gas. We were all shocked: although

* Sixty-five percent of the filed applications were returned with requests for additional information; the rest were denied.

surveying the lease area is not illegal, everyone had been under the impression that no further oil- or gas-related work had been undertaken in MC252 since the well was sealed. After my return to shore, I confirmed that oil and gas companies were again exploring the site and would soon begin drilling.

It is not surprising, of course, that oil companies remain interested in the site. After all, the Macondo well was estimated to contain as many as 1 billion barrels of oil, which could be worth $50 billion. I later confirmed with the Department of the Interior's Bureau of Ocean Energy Management (BOEM) that WesternGeco, a subsidiary of Schlumberger, one of the world's largest oil-services companies, had received a permit in October 2013 authorizing "geophysical exploration for mineral resources" in MC252. WesternGeco began its survey in January 2014. The cables we saw in April were most likely used to generate a visual representation of the area's oil and natural-gas potential.

In May 2014 BOEM approved the division of MC252 into two lease areas. BP retains just 270 acres, composed primarily of the Macondo well site and the Deepwater Horizon wreckage. The rest of the area, 5,490 acres, was turned over as a new lease to LLOG Bluewater Holdings, a privately owned offshore-oil company that is already active in the Gulf. BP cannot develop its area for oil or gas, but LLOG is allowed to drill in its lease area. Last October LLOG received approval for two exploratory wells in MC252. They will be less than a mile from the Macondo site. When I contacted LLOG in April, the vice president of deepwater projects said that the company had no immediate plans for surface operations in MC252. But he did not dispute that the company intends to recover oil from the lease block. In a March filing, LLOG outlined a plan to dig from an adjacent lease block to reach the oil in MC252. The company is expected to complete its first such well this month.

The oil industry would like us to believe that it has corrected the problems that led to the Deepwater Horizon disaster, but in June of last year the U.S. Chemical Safety Board, an independent agency established by Congress to investigate industrial incidents, released a report indicating that another "catastrophic accident" remains possible. "People in the industry have to recognize that Macondo was not just a one-off," said Cheryl MacKenzie, a lead investigator at the CSB. Virtually every investigation into the disaster has found that it resulted from industry-wide, not company-specific, failures. Yet only a handful of industry policies have changed. William K. Reilly, an EPA administrator under George H. W. Bush and the cochair of a national commission on the BP spill, warns that even good regulation and oversight cannot prevent another disaster from happening. "Drilling in very deep water is a highly challenging affair that involves highly complex technologies, and they sometimes fail," he told me. "One should not suffer the delusion that it can be done risk-free."

Nonetheless, the number of rigs operating in the deep waters of the Gulf has continued to increase—from 29 in 2011 to 51 in 2014. The amount of oil under the Gulf is estimated to be 5 billion barrels, worth about $250 billion at today's prices. Oil production there is expected to increase by 30 percent this year—meaning that by the end of 2016 oil companies will be extracting as much as 80 million gallons of oil from the Gulf each day. To reach that oil, they have begun drilling at sites nearly twice as deep as Macondo. "We're fully back in," Richard Morrison, the regional president of BP's Gulf of Mexico business, recently told a Houston energy newspaper.

Ten days after my dive, a 38-foot twin-diesel catamaran traveled seven hours from Cocodrie, Louisiana, to meet the *Atlantis* at sea. It brought us fresh honeydew melon, zucchini, eggplant, and parsley, as well as a crew member who was returning from the hospital after falling about 20 feet on his back during a maintenance operation on the *Atlantis* the previous week.

After the supplies were unloaded, I boarded the catamaran for the return trip to shore. As we glided toward Cocodrie, deep-sea oil rigs gave way to drilling platforms that proliferated so thickly that the captain had to weave around them. The contrast with Mississippi's open water was stark: Louisiana's coast is packed with rickety and abandoned oil platforms, many of which have been there for decades, the oil they sought long since gone.

Cocodrie is a tiny fishing community perched on the southern edge of Louisiana—just a few hundred homes stretching along the bayou. There are a few restaurants and one or two shops that cater to visiting sport fishers and the cannery workers who commute from farther up U.S. 56. After disembarking, I visited a grocery store that doubled as a knickknack shop. The store, owned by Cecil and Etta Lapeyrouse, was built by Cecil's grandfather in 1914.

A gas pump sat out front, and in back, hidden behind a garden of flowering white oleander and green cypress trees, a path opened onto a wooden deck drenched in sunlight, where boats could pull up to refuel. There were none in sight that day. The town's economy had suffered after BP's oil covered the water and shores near Cocodrie. Things were turning around, Cecil told me, but with the drop in business he wasn't sure whether his store would survive.

Sitting on the deck behind the shop, I admired the beauty of the empty water, now unblemished by human activity. But I looked out to sea knowing, better than ever, what lay below the placid surface.

ALEXANDRA KLEEMAN

The Bed-Rest Hoax

FROM *Harper's Magazine*

AFTER JUST A COUPLE of days on bed rest, the material of your body begins to feel different: softer, heavier, a burden to the bone beneath. The thud of the heart in the chest feels deeper: each beat shifts your frame a little. Even though you haven't used your back for anything, it aches—and when you twist into a new position, the ache swivels along with the muscles, can't be left behind. You fall asleep throughout the day but can't sleep through the night, and when you bend a limb at the joint, it's not the transparent sensation you're used to—you can feel the muscles tugging, the socket creaking in protest. Your body becomes more present, weaker, and more vulnerable: you are aware of it as though it were an alarm that has not yet gone off but could at any moment.

This summer, I checked myself into a progressive Catholic convent in the Pacific Northwest to observe the effects of five days of bed rest on my body and mind. My plan was to spend all but 30 minutes of each day in a small room with framed Bible verses on the walls, lying on my back or side on a spartan twin-size cot. In the 30 minutes I was allowed out of bed, I would shower, take bathroom breaks, or fetch food from the communal kitchen to bring back and eat in bed. In the final moments before my experiment began, I stretched the inner muscles of my thighs and blinked in the warm sunlight. I tried to take pleasure in feeling ordinary, normal, mobile.

Though five days is a relatively short bed-rest regimen, the first week is when some of the most dramatic changes to the body occur. Deconditioning of the cardiovascular system begins within 48 hours. The amount of circulating blood decreases, the heart's total output drops, and the body uses less and less oxygen. Within five days of immobilization, the arteries narrow and stiffen, and the interior lining of the blood vessels becomes less able to flex and tighten.

The body scales itself down rapidly to meet the reduced physiological demands of its new state and then pauses. Eventually, over weeks, bone density decreases and muscle volume declines. Actin and myosin, the proteins that make up muscle, break down into free-floating nitrogen that is flushed from the body through the kidneys. Simply standing up can cause fainting, since the body is no longer used to pumping blood against the pull of gravity.

Hundreds of thousands of years of evolution have enabled us to walk upright, a task few other mammals can manage—sheep and rabbits often lose consciousness or die when held vertical. But the more time a body spends away from plumb, the greater its difficulty in readapting to normal life. For this reason, bed rest is used as an analog for space travel in NASA experiments: the effect of weightlessness on human bodies can be simulated on earth by putting subjects to bed at a six-degree negative incline. Prolonged rest is an extreme physiological challenge, a new environment for the body to navigate.

What I've described sounds like a sort of bodily erosion, a slow injury or gentle decay, but it also happens to be one of the most commonly prescribed treatments in the United States for pregnant women at risk of preterm birth. Each year as many as 700,000 pregnant women are prescribed some form of bed rest: from several hours a day to round-the-clock immobilization with breaks only to use the bathroom. For some types of high-risk pregnancy, the mother-to-be is hospitalized and prohibited from getting up to relieve or clean herself, from standing, or even from sitting propped up in bed. Strict bed rest—whether at home or in a hospital—often means that a woman has to forfeit exercise, income, and normal domestic tasks such as caring for her family or maintaining her home.

The practice continues despite a growing body of clinical evidence showing that strict bed rest offers no benefits to the fetus or to the mother. It has not been proved effective in treating gestational hypertension, preeclampsia, a shortened cervix, spontaneous abortion, or impaired fetal growth. The hazards of bed rest, on the other hand, are well substantiated: patients may suffer from bone loss, blood clots, muscle atrophy, weight loss, and psychological malaise. Enrollment in one study, in which women carrying twins were randomly admitted to the hospital for bed rest or assigned outpatient care with no activity restriction, was halted midway because of concerns about a possible detrimental effect to the hospitalized group.

Even so, bed rest remains a routine therapeutic intervention for pregnancy, with up to 95 percent of obstetricians reporting that they've prescribed it for their patients. Decades after the treatment fell out of favor for other conditions, pregnancy is the last remaining medical territory to which bed rest can lay claim. It is now the domain of those physically incapable of movement—those, for example, who have broken all their limbs—and expectant mothers.

When John Hilton published *Rest and Pain,* his influential 1863 treatise on the beneficial effects of rest, he was writing for an audience that was generally suspicious of the idea of taking to bed. The hospital ward in particular was seen as synonymous with death, in part because of the ease with which infections spread from patient to patient before sanitation standards were adopted.

Hilton sought to change that perception. He argued that nature was the primary agent of healing and that the physician's best course of action was to let the body rest, so that it might heal itself. The physician could be seen as nature's assistant, a helpful nurse: "In fact," wrote Hilton, "nearly all our best considered operations are done for the purpose of making it possible to keep the structures at rest, or freeing Nature from the disturbing cause which was exhausting her powers, or making her repeated attempts at repair unavailing."

Physicians took Hilton's recommendations to heart, and rest became the guiding principle of medical interventions, leaving nurses responsible only for the maintenance of good hygiene and the prevention of bedsores. Soon, myocardial infarction, congestive heart failure, tuberculosis, peptic ulcers, and rheumatic fever were all being treated with bed rest. Because rest was an unlimited good, patients were often put to bed at home or in the hospital for indefinite periods of time—the longer the better.

One of the most well respected of these therapies was a "rest cure" that was developed by Silas Weir Mitchell, a physician and author, to treat neurasthenia, a bundle of physical and psychological symptoms that we might diagnose today as depression or anxiety. One monograph on neurasthenia contains a list of 81 symptoms, including insomnia, bad dreams, mental irritation, rapid decay of the teeth, dizziness, hopelessness, deficient thirst, vague pains, vertigo, cold hands and feet, and "fear of everything"—a list that the author admits is "not exhaustive." Women were especially susceptible to neurasthenia, Mitchell wrote, above all "nervous women, who, as a rule, are thin and lack blood." Their bodies were continually in flux, passing from puberty to pregnancy to menopause, and so were an inherent source of destabilization and pathology.

Women who suffered from nervous pathologies were isolated from friends and family and confined to bed for weeks at a time. In the beginning stages of treatment, patients were forbidden to sit up, sew, read, write, or use their hands for any activity except cleaning the teeth. Each day involved a regimen of "passive exercise," which consisted of full-body massage and electrical stimulation of the muscles. To counteract the loss of body mass, women were fed a diet that started with a week on an all-milk regimen; patients were conditioned to consume two quarts a day. Over time, they worked up to rich meals comprising mutton chops, cod-liver oil, malt extract, more milk, and doses of a raw-beef soup that was made by dissolving meat with a few drops of hydrochloric acid.

The aim was to produce a more resilient woman by cushioning her systems with blood and fat, and to make her psyche resemble the stillness of her outer flesh at rest rather than the mercurial, reactive processes of the womb. But there was also a punitive dimension to Mitchell's treatment: he believed that his weak-nerved patients had been coddled by those around them. The neurasthenic woman was "a vampire who sucks the blood of the healthy people about her," her morality spoiled by undisciplined care and concern. "The moral uses of enforced rest are readily estimated," Mitchell writes. "From a restless life of irregular hours, and probably endless drugging, from hurtful sympathy and overzealous care, the patient passes to an atmosphere of quiet, to order and control, to the system and care of a thorough nurse, to an absence of drugs, and to simple diet." Mitchell directed women who had lived selfishly, governed by concern for their own well-being and mental life, to turn their thoughts away from their condition and to focus instead on their duty to others.

This brand of paternalism has mostly disappeared from modern medicine, but its vestiges can be seen in the way we care for pregnant women, whose perceived selfishness (the impulse to continue working or to have a cup of coffee or a glass of wine) is often cast as a threat to their unborn children. When other branches of medicine have abandoned bed rest as a therapeutic tool, why does it linger on in prenatal care? Maybe the answer has to do with the hold that a particular kind of androcentric worldview has over women's bodies. Though men and women are both made of flesh, women have long been viewed as the fleshier sex, their mental processes unavoidably interwoven with those of their reproductive organs. But even though women were understood to be controlled by their bodies, they were paradoxically capable of obstructing the body's natural order by exercising autonomy—which could mean deciding not to bear children or threatening gestation through excessive activity and worry. Meanwhile, after doctors observed that wounded veterans returning from World War II recovered more completely from their injuries when they spent less time confined to bed, the treatment was essentially abandoned for male patients.

One of the best-known fictional treatments of Mitchell's rest cure is Charlotte Perkins Gilman's short story "The Yellow Wallpaper," about a woman who is confined to bed by her doctor husband and forbidden intellectual stimulation. Gilman was a patient of Mitchell's, and she spent a month at his clinic. When he sent her home, he instructed her to live "as domestic a life as possible," lying down after every meal, restricting intellectual activity to a maximum of two hours a day, and heeding his warning to "never touch pen, brush, or pencil as long as you live." Under Mitchell's instructions, Gilman's mental agony only increased, a "mental torment . . . so heavy in its nightmare gloom that it seemed real enough to dodge." At the end of "The Yellow Wallpaper," Gilman's protagonist goes insane.

On message boards and in chatrooms, mothers with high-risk pregnancies convene to trade advice regarding bed rest. Women compare the amount of bed rest prescribed (I saw a range from 2 weeks to 25), ask one another for clarification of their doctors' orders (are you allowed to sit up?), give practical suggestions (get a minifridge to put by your bed), and discuss ways to pass the time (coloring books, puzzles, Hulu, crocheting, knitting, journaling, posting on Internet message boards). They trade tips on how to reduce back pain, leg cramps, and numbness in the extremities. Messages are supportive and punctuated by smiley faces and small pixelated images of flowers. Below every post on BabyCenter's Bed Rest Club forum is a button that allows you to send the writer a virtual hug.

"Does anyone else have days where they just want to cry?" one post asks. "Anyone get put on bed rest and lose all income," reads another. A strain of guilt and self-recrimination runs through many of the messages: "Being told I had to stop working was really hard, being told I was on bedrest was hard, but being told that my body is failing my baby, that's the worst. I haven't even started being a mother and I already feel like a failure."

For mothers struggling with the effect of bed rest on their families, on their finances, and on their own mental health, adhering strictly to the obstetrician's orders can serve as an antidote to feelings of powerlessness, a doctor-approved avenue through which the mother's will can be exercised over her own body. A difficult pregnancy can be transformed into a task that is worked day by day, and online message chains are filled with reminders that the discomfort and stress of being bedridden will all be worth it once the baby has been born healthy. A post titled "Success Story" concludes: "At 24 weeks I was told my cervix was at a 1.1 and had started dilating and funneling. I was put on bed rest with progesterone suppositories . . . I stayed in bed and only got up to pee and shower. I made it to 39 weeks and actually had to be induced. You can do this ladies. My bed rest baby is now 2 years old."

These testimonials motivate bedridden mothers to keep going, to believe in their own ability to change the course of their pregnancy, and to "keep that baby cooking!" But the logic of the community is self-reinforcing: almost all the women posting on the site are on bed rest or have been in the past, and successful births are retroactively cast as a direct consequence of the time spent in bed. Counterexamples are vanishingly rare, as are community members who've ignored a recommendation to go on bed rest—though there are some who have persuaded their doctors to prescribe the treatment after reading in online forums about its success. (Community members likewise talk about successful interventions with the drug terbutaline, which the FDA warned in 2011 "should not be used for prevention or prolonged treatment . . . of preterm labor.") Failing to follow the guidelines of her physician, or her own sometimes-stricter vision of how much movement she can afford to inflict on her womb, can make a woman with a high-risk pregnancy worry that she has harmed her baby. Contractions or tenderness that follow a day during which she got up or walked more often than she feels she should have are easily perceived as a consequence of her own carelessness or neglect.

Clinical trials are rarely cited on bed-rest message boards, though the moderator of one forum sent me a paper published in 2015 in the *American Journal of Health Economics* that claimed to show a decrease in very low birth weights and very premature outcomes. It was a statistical analysis of survey responses rather than a controlled trial, and it made no distinction between patients on bed rest for two days and patients on bed rest for weeks or months at a time. I found two small trials, conducted in 1983 and 1992, that suggested there might be some benefit for patients with hypertension. The 1983 trial, however, also noted that a more moderate prescription of four to six hours of rest a day is equally effective in lowering blood pressure.

These studies contrast with the Cochrane reviews of bed rest, which represent the most comprehensive assessment of the available science, and draw from dozens of peer-reviewed papers. Those have consistently shown no proven benefit from the treatment and do not recommend it. A 2013 study of pregnant women with short cervixes found that preterm birth was more likely among those placed on activity restriction. Other studies have shown that pregnant women are particularly vulnerable to the negative side effects of bed rest: an increase in clotting factors during pregnancy makes patients more likely to form blood clots, and immobilization compresses the veins further, putting patients at even greater risk. (Pulmonary embolism is the cause of 10 percent of pregnancy-related deaths.)

Obstetricians who have research experience are far less likely to recommend bed rest than those who do not. "Out in the community, you're going to have doctors that say, 'Absolutely, you should be on bed rest,'" said Cynthia Gyamfi-Bannerman, an associate professor of obstetrics and gynecology at Columbia University Medical Center, where the staff actively discourages the practice. "One of the most common questions I get from my pregnant patients is, 'When am I going to be on bed rest?' We tell them, well, hopefully, never. It's harder, almost, to say, 'You don't need it,' than to say, 'OK, sure, go ahead.'"

Most of the Cochrane reviews on bed rest were published within the past 10 years. Christina Herrera, who is a fellow in maternal-fetal medicine at the University of Utah, said that she learned about the general complications of being confined to bed in medical school but didn't encounter clinical research about the risks of the practice for pregnancy until her residency. Herrera now tells her patients to avoid strict bed rest at all costs.

One of the greatest obstacles to changing the way bed rest is prescribed is the therapeutic imperative. "Providers feel that they have to do something to help the pregnancy," Herrera said, "even if there's nothing they can do. And so women of course jump at the chance to do anything they can . . . if it'll benefit the baby in any way, shape, or form." She added: "Historically, women are sacrificial."

Through some combination of ignorance and wishful thinking, bed rest survives. In 2009 a young woman named Samantha Burton experienced symptoms of preterm labor 15 weeks before her due date and went voluntarily to Tallahassee Memorial Hospital. She was seen by a doctor who told her that she would have to be admitted and remain in bed. Burton, who had two small children, agreed to rest but wanted to go home. She also wanted a second opinion. The doctor told her that she would not be allowed to leave and initiated legal proceedings to confine her to the hospital. A judge found in favor of Tallahassee Memorial and issued a court order mandating hospital bed rest, medication to prolong her pregnancy, and, if necessary, forced delivery. Three days later, Burton delivered a stillborn baby by cesarean section.

My first night in the convent, I had a quick dinner of scrambled eggs and bagels with the nuns and oblates in the cafeteria. They were kind enough to let me undergo my bed-rest experiment on their grounds, in a wing of the complex meant for spiritual retreatants—all of whom had taken vows of silence for the duration of their stay there, except for me. The nuns were extremely kind and friendly, and unabashedly curious about me, the only visitor to their retreat center who was allowed to talk to them. They asked me about bed rest, why I was doing it, what I thought I would find out. They introduced me to the oldest sister, who had been a nun for more than 70 years. They thought she might be able to tell me about a time when people took to bed more often. (She couldn't.) They wished me mental and spiritual peace even though they knew that was not what I was there for. Before I left the table, a nun blessed me, blessed my article, and blessed my writing. These were the last people I'd speak to face-to-face for the next five days.

I told the nuns that I expected my time on bed rest to make me better rested, and probably very bored. At that point, I had no idea how draining it would be to adjust my body continuously, as one part after another complained about each new position. I lay on my back or on my side, watching a small rectangle of light blaze and flicker in the afternoon, dimming as the day turned to evening. I breathed deeply even though it felt like my lungs were trapped beneath an invisible new weight. I learned to lie on my side with one leg straight and one bent to avoid the pain of my knee digging into the increasingly tender flesh of the leg beneath. My whole body felt weak and sore and sensitive by day two—but I still felt generally healthy, as though I could shake off this new frailty if I were just allowed to go for a long walk.

By the third day, I found myself thinking that something had gone wrong inside my body, that there was something besides the experiment harming me. What else could explain how unwell I felt? The muscles of my neck and shoulders were alternately sore and numb; my legs hurt when I rested them against each other and ached when they were apart; my heart raced when I turned from one side to the other. I felt sad for no good reason, unfocused but undistractible—I couldn't seem to get my mind off how I felt, but I was unable to bring the experience into sharp-enough relief to analyze it. It's more difficult to think when you're horizontal: alertness comes in plains and troughs rather than in peaks. None of my thoughts had any lift to them.

The few times a day when I let myself out of bed, I could feel how different my body had become. Standing up made me dizzy and set my heart pounding. Walking 20 paces to the kitchen exhausted me and left me feeling empty, my legs shaking. I felt as though I'd ended up on the wrong planet, or in the wrong body. I knew from reading NASA's bed-rest studies that the first week is supposed to be the worst, full of strange pains and headaches, urgent discomforts that last for hours and then fade away. I didn't know if the second week is easier because the body has adapted, or whether after a week in bed it just gets harder to remember what being out of bed used to feel like, what having a body used to feel like. By the fifth day, I felt as though I'd aged 20 years.

I had read papers that described the dramatic effect of immobilization on the human psyche. Adolescent girls in full-body casts viewed the motion of others with jealousy, tracking them with their eyes, simulating movement with twitches and fidgeting. Individuals forced to rest acted out emotionally, manifesting fear, guilt, and anger. Immobilization altered the perception of weight, pressure, temperature, pattern, and form, and it distorted the experience of time.

The day I left the convent, I wheeled my small, heavy rolling luggage back up the road that led to the bus stop. The distance seemed to have dilated, the inclines and slopes had become steeper. A walk that had taken me 10 minutes on the first day now took me closer to 30.

Lying in bed on my last day of rest, I thought about the phrase "First, do no harm," which is commonly believed to be part of the Hippocratic oath even though it appeared much later. Intuitively, bed rest feels like it should be a harmless therapy: What danger could there be in doing, for an exaggerated period, something we do every night? If there's no benefit in it, at least there should be no harm. But in every other branch of medicine, we insist that a treatment justify its use through empirical evidence, through proof of its capacity to help. There's no excuse for letting any patient suffer the known harm of immobilization without compelling evidence of its benefits. We should recognize bed rest for what it is: not an escape from risk but the adoption of a new risk, one whose outcome is uncertain, but not unknown.

ELIZABETH KOLBERT

The Siege of Miami

FROM *The New Yorker*

THE CITY OF MIAMI BEACH floods on such a predictable basis that if, out of curiosity or sheer perversity, a person wants to she can plan a visit to coincide with an inundation. Knowing the tides would be high around the time of the "super blood moon," in late September, I arranged to meet up with Hal Wanless, the chairman of the University of Miami's Geological Science Department. Wanless, who is 73, has spent nearly half a century studying how South Florida came into being. From this, he's concluded that much of the region may have less than half a century more to go.

We had breakfast at a greasy spoon not far from Wanless's office, then set off across the MacArthur Causeway. (Out-of-towners often assume that Miami Beach is part of Miami, but it's situated on a separate island, a few miles off the coast.) It was a hot, breathless day, with a brilliant blue sky. Wanless turned onto a side street, and soon we were confronting a pond-sized puddle. Water gushed down the road and into an underground garage. We stopped in front of a four-story apartment building, which was surrounded by a groomed lawn. Water seemed to be bubbling out of the turf. Wanless took off his shoes and socks and pulled on a pair of polypropylene booties. As he stepped out of the car, a woman rushed over. She asked if he worked for the city. He said he did not, an answer that seemed to disappoint but not deter her. She gestured at a palm tree that was sticking out of the drowned grass.

"Look at our yard, at the landscaping," she said. "That palm tree was superexpensive." She went on, "It's crazy—this is salt water."

"Welcome to rising sea levels," Wanless told her.

According to the Intergovernmental Panel on Climate Change, sea levels could rise by more than three feet by the end of this century. The United States Army Corps of Engineers projects that they could rise by as much as five feet; the National Oceanic and Atmospheric Administration predicts up to six and a half feet. According to Wanless, all these projections are probably low. In his office, Wanless keeps a jar of meltwater he collected from the Greenland ice sheet. He likes to point out that there is plenty more where that came from.

"Many geologists, we're looking at the possibility of a ten-to-thirty-foot range by the end of the century," he told me.

We got back into the car. Driving with one hand, Wanless shot pictures out the window with the other. "Look at that," he said. "Oh, my gosh!" We'd come to a neighborhood of multimillion-dollar homes where the water was creeping under the security gates and up the driveways. Porsches and Mercedeses sat flooded up to their chassis.

"This is today, you know," Wanless said. "This isn't with two feet of sea level rise." He wanted to get better photos, and pulled over onto another side street. He handed me the camera so that I could take a picture of him standing in the middle of the submerged road. Wanless stretched out his arms, like a magician who'd just conjured a rabbit. Some workmen came bouncing along in the back of a pickup. Every few feet, they stuck a depth gauge into the water. A truck from the Miami Beach Public Works Department pulled up. The driver asked if we had called City Hall. Apparently, one of the residents of the street had mistaken the high tide for a water main break. As we were chatting with him, an elderly woman leaning on a walker rounded the corner. She looked at the lake the street had become and wailed, "What am I supposed to do?" The men in the pickup truck agreed to take her home. They folded up her walker and hoisted her into the cab.

To cope with its recurrent flooding, Miami Beach has already spent something like a hundred million dollars. It is planning on spending several hundred million more. Such efforts are, in Wanless's view, so much money down the drain. Sooner or later—and probably sooner—the city will have too much water to deal with. Even before that happens, Wanless believes, insurers will stop selling policies on the luxury condos that line Biscayne Bay. Banks will stop writing mortgages.

"If we don't plan for this," he told me, once we were in the car again, driving toward the Fontainebleau hotel, "these are the new Okies." I tried to imagine Ma and Pa Joad heading north, their golf bags and espresso machine strapped to the Range Rover.

The amount of water on the planet is fixed (and has been for billions of years). Its distribution, however, is subject to all sorts of rearrangements. In the coldest part of the last ice age, about 20,000 years ago, so much water was tied up in ice sheets that sea levels were almost 400 feet lower than they are today. At that point, Miami Beach, instead of being an island, was 15 miles from the Atlantic coast. Sarasota was 100 miles inland from the Gulf of Mexico, and the outline of the Sunshine State looked less like a skinny finger than like a plump heel.

As the ice age ended and the planet warmed, the world's coastlines assumed their present configuration. There's a good deal of evidence—much of it now submerged—that this process did not take place slowly and steadily but, rather, in fits and starts. Beginning around 12,500 BC, during an event known as meltwater pulse 1A, sea levels rose by roughly 50 feet in three or four centuries, a rate of more than a foot per decade. Meltwater pulse 1A, along with pulses 1B, 1C, and 1D, was, most probably, the result of ice sheet collapse. One after another, the enormous glaciers disintegrated and dumped their contents into the oceans. It's been speculated—though the evidence is sketchy—that a sudden flooding of the Black Sea toward the end of meltwater pulse 1C, around 7,500 years ago, inspired the deluge story in Genesis.

As temperatures climb again, so, too, will sea levels. One reason for this is that water, as it heats up, expands. The process of thermal expansion follows well-known physical laws, and its impact is relatively easy to calculate. It is more difficult to predict how the earth's remaining ice sheets will behave, and this difficulty accounts for the wide range in projections.

Low-end forecasts, like the IPCC's, assume that the contribution from the ice sheets will remain relatively stable through the end of the century. High-end projections, like NOAA's, assume that ice melt will accelerate as the earth warms (as, under any remotely plausible scenario, the planet will continue to do at least through the end of this century, and probably beyond). Recent observations, meanwhile, tend to support the most worrisome scenarios.

The latest data from the Arctic, gathered by a pair of exquisitely sensitive satellites, show that in the past decade Greenland has been losing more ice each year. In August NASA announced that, to supplement the satellites, it was launching a new monitoring program called—provocatively—Oceans Melting Greenland, or OMG. In November researchers reported that, owing to the loss of an ice shelf off northeastern Greenland, a new "floodgate" on the ice sheet had opened. All told, Greenland's ice holds enough water to raise global sea levels by 20 feet.

At the opposite end of the earth, two groups of researchers—one from NASA's Jet Propulsion Lab and the other from the University of Washington—concluded last year that a segment of the West Antarctic ice sheet has gone into "irreversible decline." The segment, known as the Amundsen Sea sector, contains enough water to raise global sea levels by 4 feet, and its melting could destabilize other parts of the ice sheet, which hold enough ice to add 10 more feet. While the "decline" could take centuries, it's also possible that it could be accomplished a lot sooner. NASA is already planning for the day when parts of the Kennedy Space Center, on Florida's Cape Canaveral, will be underwater.

The day I toured Miami Beach with Hal Wanless, I also attended a panel discussion at the city's Convention Center titled "Eyes on the Rise." The discussion was hosted by the French government, as part of the lead-up to the climate convention in Paris, at that point two months away. Among the members of the panel was a French scientist named Eric Rignot, a professor at the University of California, Irvine. Rignot is one of the researchers on OMG, and in a conference call with reporters during the summer he said he was "in awe" of how fast the Greenland ice sheet was changing. I ran into him just as he was about to go onstage.

"I'm going to scare people out of this room," he told me. His fellow panelists were a French geophysicist, a climate scientist from the University of Miami, and Miami Beach's mayor, Philip Levine. Levine was elected in 2013, after airing a commercial that tapped into voters' frustration with the continual flooding. It showed him preparing to paddle home from work in a kayak.

"Some people get swept into office," Levine joked when it was his turn at the mike. "I always say I got floated in." He described the steps his administration was taking to combat the effects of rising seas. These include installing enormous underground pumps that will suck water off the streets and dump it into Biscayne Bay. Six pumps have been completed, and 54 more are planned. "We had to raise people's storm-water fees to be able to pay for the first hundred-million-dollar tranche," Levine said. "So picture this: you get elected to office and the first thing you tell people is, 'By the way, I'm going to raise your rates.' "

He went on, "When you are doing this, there's no textbooks, there's no 'How to Protect Your City from Sea Level Rise,' go to Chapter Four." So the city would have to write its own. "We have a team that's going to get it done, that's going to protect this city," the mayor said. "We can't let investor confidence, resident confidence, confidence in our economy start to fall away."

John Morales, the chief meteorologist at NBC's South Florida affiliate, was moderating the discussion. He challenged the mayor, offering a version of the argument I'd heard from Wanless—that today's pumps will be submerged by the seas of tomorrow.

"Down the road, this is just a Band-Aid," Morales said.

"I believe in human innovation," Levine responded. "If, thirty or forty years ago, I'd told you that you were going to be able to communicate with your friends around the world by looking at your watch or with an iPad or an iPhone, you would think I was out of my mind." Thirty or 40 years from now, he said, "We're going to have innovative solutions to fight back against sea level rise that we cannot even imagine today."

Many of the world's largest cities sit along a coast, and all of them are, to one degree or another, threatened by rising seas. Entire countries are endangered—the Maldives, for instance, and the Marshall Islands. Globally, it's estimated that a hundred million people live within three feet of mean high tide and another hundred million or so live within six feet of it. Hundreds of millions more live in areas likely to be affected by increasingly destructive storm surges.

Against this backdrop, South Florida still stands out. The region has been called "ground zero when it comes to sea level rise." It has also been described as "the poster child for the impacts of climate change," the "epicenter for studying the effects of sea level rise," a "disaster scenario," and "the New Atlantis." Of all the world's cities, Miami ranks second in terms of assets vulnerable to rising seas—number one is Guangzhou—and in terms of population it ranks fourth, after Guangzhou, Mumbai, and Shanghai. A recent report on storm surges in the United States listed four Florida cities among the eight most at risk. (On that list, Tampa came in at number one.) For the past several years, the daily high-water mark in the Miami area has been racing up at the rate of almost an inch a year, nearly 10 times the rate of average global sea-level rise. It's unclear exactly why this is happening, but it's been speculated that it has to do with changes in ocean currents which are causing water to pile up along the coast. Talking about climate change in the Everglades this past Earth Day, President Obama said, "Nowhere is it going to have a bigger impact than here in South Florida."

The region's troubles start with its topography. Driving across South Florida is like driving across central Kansas, except that South Florida is greener and a whole lot lower. In Miami-Dade County, the average elevation is just 6 feet above sea level. The county's highest point, aside from manmade structures, is only about 25 feet, and no one seems entirely sure where it is. (The humorist Dave Barry once set out to climb Miami-Dade's tallest mountain, and ended up atop a local garbage dump nicknamed Mount Trashmore.) Broward County, which includes Fort Lauderdale, is equally flat and low, and Monroe County, which includes the Florida Keys, is even more so.

But South Florida's problems also run deeper. The whole region—indeed, most of the state—consists of limestone that was laid down over the millions of years Florida sat at the bottom of a shallow sea. The limestone is filled with holes, and the holes are, for the most part, filled with water. (Near the surface, this is generally fresh water, which has a lower density than salt water.)

Until the 1880s, when the first channels were cut through the region by steam-powered dredges, South Florida was one continuous wetland—the Everglades. Early efforts to drain the area were only half successful; northerners lured by turn-of-the-century real-estate scams found the supposedly rich farmland they'd purchased was more suitable for swimming.

"I have bought land by the acre, and I have bought land by the foot; but, by God, I have never before bought land by the gallon," one arrival from Iowa complained.

Even today, with the Everglades reduced to half its former size, water in the region is constantly being shunted around. The South

Florida Water Management District, a state agency, claims that it operates the "world's largest water control system," which includes 2,300 miles of canals, 61 pump stations, and more than 2,000 "water control structures." Floridians south of Orlando depend on this system to prevent their lawns from drowning and their front steps from becoming docks. (Basement flooding isn't an issue in South Florida, because no one has a basement—the water table is too high.)

When the system was designed—redesigned, really—in the 1950s, the water level in the canals could be maintained at least a foot and a half higher than the level of high tide. Thanks to this difference in elevation, water flowed off the land toward the sea. At the same time, there was enough fresh water pushing out to prevent salt water from pressing in. Owing in part to sea level rise, the gap has since been cut by about eight inches, and the region faces the discomfiting prospect that, during storms, it will be inundated not just along the coasts but also inland, by rainwater that has nowhere to go. Researchers at Florida Atlantic University have found that with just six more inches of sea level rise the district will lose almost half its flood-control capacity. Meanwhile, what's known as the saltwater front is advancing. One city—Hallandale Beach, just north of Miami—has already had to close most of its drinking wells, because the water is too salty. Many other cities are worried that they will have to do the same.

Jayantha Obeysekera is the Water Management District's chief modeler, which means it's his job to foresee South Florida's future. One morning, I caught up with him at a flood-control structure known as S13, which sits on a canal known as C11, west of Fort Lauderdale.

"We have a triple whammy," he said. "One whammy is sea level rise. Another whammy is the water table comes up higher, too. And in this area the higher the water table, the less space you have to absorb storm water. The third whammy is if the rainfall extremes change, and become more extreme. There are other whammies probably that I haven't mentioned. Someone said the other day, 'The water comes from six sides in Florida.'"

A month after the super blood moon, South Florida experienced another series of very high tides—"king tides," as Miamians call them. This time, I went out to see the effects with Nicole Hernandez Hammer, an environmental-studies researcher who works for the Union of Concerned Scientists. Hammer had looked over elevation maps and decided that Shorecrest, about five miles north of downtown Miami, was a neighborhood where we were likely to find flooding. It was another hot, blue morning, and as we drove along, in Hammer's Honda, at first it seemed that she'd miscalculated. Then, all of a sudden, we arrived at a major intersection that was submerged. We parked and made our way onto a side street, also submerged. We were standing in front of a low- slung apartment building, debating what to do next, when one of the residents came by.

"I've been trying to figure out: Where is the water coming from?" he said. "It'll be drying up and then it'll be just like this again." He had complained to the building's superintendent. "I told him, 'Something needs to be done about this water, man.' He says he'll try to do something." A cable-repair truck trailing a large wake rolled by and then stalled out.

The water on the street was so deep that it was, indeed, hard to tell where it was coming from. Hammer explained that it was emerging from the storm drains. Instead of funneling rainwater into the bay, as they were designed to do, the drains were directing water from the bay onto the streets. "The infrastructure we have is built for a world that doesn't exist anymore," she said.

Neither of us was wearing boots, a fact that, as we picked our way along, we agreed we regretted. I couldn't help recalling stories I'd heard about Miami's antiquated sewer system, which leaks so much raw waste that it's the subject of frequent lawsuits. (To settle a suit brought by the federal government, the county recently agreed to spend $1.6 billion to upgrade the system, though many question whether the planned repairs adequately account for sea level rise.) Across the soaked intersection, in front of a single-family home, a middle-aged man was unloading groceries from his car. He, too, told us he didn't know where the water was coming from.

"I heard on the news it's because the moon turned red," he said. "I don't have that much detail about it." During the past month, he added, "it's happened very often." (In an ominous development, Miami this past fall experienced several very high tides at times of the month when, astronomically speaking, it shouldn't have.)

"Honestly, sometimes, when I'm talking to people, I think, *Oh,*

I wish I had taken more psychology courses," Hammer told me. A lot of her job involves visiting low-lying neighborhoods like Shorecrest, helping people understand what they're seeing. She shows them elevation maps and climate-change projections, and explains that the situation is only going to get worse. Often, Hammer said, she feels like a doctor: "You hear that they're trying to teach these skills in medical schools, to encourage them to have a better bedside manner. I think I might try to get that kind of training, because it's really hard to break bad news."

It was garbage-collection day, and in front of one house county-issued trash bins bobbed in a stretch of water streaked with oil. Two young women were surveying the scene from the driveway, as if from a pier.

"It's horrible," one of them said to us. "Sometimes the water actually smells." They were sisters, originally from Colombia. They wanted to sell the house, but, as the other sister observed, "No one's going to want to buy it like this."

"I have called the City of Miami," the first sister said. "And they said it's just the moon. But I don't think it's the moon anymore."

After a couple of minutes, their mother came out. Hammer, who was born in Guatemala, began chatting with her in Spanish. "Oh," I heard the mother exclaim. "Dios mío! El cambio climático!"

Marco Rubio, Florida's junior senator, who has been running third in Republican primary polls, grew up not far from Shorecrest, in West Miami, which sounds like it's a neighborhood but is actually its own city. For several years, he served in Florida's House of Representatives, and his district included Miami's flood-vulnerable airport. Appearing this past spring on *Face the Nation,* Rubio was asked to explain a statement he had made about climate change. He offered the following: "What I said is, humans are not responsible for climate change in the way some of these people out there are trying to make us believe, for the following reason: I believe that climate is changing because there's never been a moment where the climate is not changing."

Around the same time, it was revealed that aides to Florida's governor, Rick Scott, also a Republican, had instructed state workers not to discuss climate change, or even to use the term. The Scott administration, according to the Florida Center for Investigative Reporting, also tried to ban talk of sea level rise; state employees were supposed to speak, instead, of "nuisance flooding." Scott denied having imposed any such Orwellian restrictions, but I met several people who told me they'd bumped up against them. One was Hammer, who, a few years ago, worked on a report to the state about threats to Florida's transportation system. She said that she was instructed to remove all climate-change references from it. "In some places, it was impossible," she recalled. "Like when we talked about the Intergovernmental Panel on Climate Change, which has 'climate change' in the title."

Scientists who study climate change (and the reporters who cover them) often speculate about when the partisan debate on the issue will end. If Florida is a guide, the answer seems to be never. During September's series of king tides, former vice president Al Gore spent a morning sloshing through the flooded streets of Miami Beach with Mayor Levine, a Democrat. I met up with Gore the following day, and he told me that the boots he'd worn had turned out to be too low; the water had poured in over the top.

"When the governor of the state is a full-out climate denier, the irony is just excruciatingly painful," Gore observed. He said that he thought Florida ought to "join with the Maldives and some of the small island states that are urging the world to adopt stronger restrictions on global-warming pollution."

Instead, the state is doing the opposite. In October Florida filed suit against the Environmental Protection Agency, seeking to block new rules aimed at limiting warming by reducing power-plant emissions. (Two dozen states are participating in the lawsuit.)

"The level of disconnect from reality is pretty profound," Jeff Goodell, a journalist who's working on a book on the impacts of sea level rise, told me. "We're sort of used to that in the climate world. But in Florida there are real consequences. The water is rising right now."

Meanwhile, people continue to flock to South Florida. Miami's metropolitan area, which includes Fort Lauderdale, has been one of the fastest growing in the country; from 2013 to 2014, in absolute terms it added more residents than San Francisco and, proportionally speaking, it outdid Los Angeles and New York. Currently, in downtown Miami there are more than 25,000 new condominium units either proposed or under construction. Much of the boom is being financed by "flight capital" from countries like Argentina and Venezuela; something like half of recent home sales in Miami were paid for in cash.

And just about everyone who can afford to buys near the water. Not long ago, Kenneth Griffin, a hedge fund billionaire, bought a penthouse in Miami Beach for $60 million, the highest amount ever paid for a single-family residence in Miami-Dade County (and $10 million more than the original asking price). The penthouse, in a new building called Faena House, offers eight bedrooms and a 70-foot rooftop pool. When I read about the sale, I plugged the building's address into a handy program called the Sea Level Rise Toolbox, created by students and professors at Florida International University. According to the program, with a little more than one foot of rise the roads around the building will frequently flood. With two feet, most of the streets will be underwater, and with three it seems that, if Faena House is still habitable, it will be accessible only by boat.

I asked everyone I met in South Florida who seemed at all concerned about sea level rise the same question: What could be done? More than a quarter of the Netherlands is below sea level and those areas are home to millions of people, so low-elevation living is certainly possible. But the geology of South Florida is peculiarly intractable. Building a dike on porous limestone is like putting a fence on top of a tunnel: it alters the route of travel, but not necessarily the amount.

"You can't build levees on the coast and stop the water" is the way Jayantha Obeysekera put it. "The water would just come underground."

Some people told me that they thought the only realistic response for South Florida was retreat.

"I live opposite a park," Philip Stoddard, the mayor of South Miami—also a city in its own right—told me. "And there's a low area in it that fills up when it rains. I was out there this morning walking my dog, and I saw fish in it. Where the heck did the fish come from? They came from underground. We have fish that travel underground!

"What that means is, there's no keeping the water out," he went on. "So ultimately this area has to depopulate. What I want to work toward is a slow and graceful depopulation, rather than a sudden and catastrophic one."

More often, I heard echoes of Mayor Levine's Apple Watch line. Who knows what amazing breakthroughs the future will bring?

"I think people are underestimating the incredible innovative imagination in the world of adaptive design," Harvey Ruvin, the clerk of the courts of Miami-Dade County and the chairman of the county's Sea Level Rise Task Force, said when I went to visit him in his office. A quote from Buckminster Fuller hung on the wall: "We are all passengers on Spaceship Earth." Ruvin became friendly with Fuller in the 1960s, after reading about a plan Fuller had drawn up for a floating city in Tokyo Bay.

"I would agree that things can't continue exactly the way they are today," Ruvin told me. "But what we will evolve to may be better."

"I keep telling people, 'This is my patient,'" Bruce Mowry, Miami Beach's city engineer, was saying. "I can't lose my patient. If I don't do anything, Miami Beach may not be here." It was yet another day of bright-blue skies and "nuisance flooding," and I was walking with Mowry through one of Miami Beach's lowest neighborhoods, Sunset Harbour.

If Miami Beach is on a gurney, then Mowry might be said to be thumping its chest. It's his job to keep the city viable, and since no one has yet come up with a smart-watch-like breakthrough, he's been forced to rely on more primitive means, like pumps and asphalt. We rounded a corner and came to a set of stairs, which led down to some restaurants and shops. Until recently, Mowry explained, the shops and the street had been at the same level. But the street had recently been raised. It was now almost a yard higher than the sidewalk.

"I call this my five-step program," he said. "What are the five steps?" He counted off the stairs as we descended: "One, two, three, four, five." Some restaurants had set up tables at the bottom, next to what used to be a curb but now, with the elevation of the road, is a three-foot wall. Cars whizzed by at the diners' eye level. I found the arrangement disconcerting, as if I'd suddenly shrunk. Mowry told me that some of the business owners, who had been unhappy when the street flooded, now were unhappy because they had no direct access to the road: "It's, like, can you win?"

Several nearby streets had also been raised, by about a foot. The elevated roadbeds were higher than the driveways, which now all sloped down. The parking lot of a car-rental agency sat in a kind of hollow.

I asked about the limestone problem. "That is the one that scares us more than anything," Mowry said. "New Orleans, the Netherlands—everybody understands putting in barriers, perimeter levees, pumps. Very few people understand: What do you do when the water's coming up through the ground?

"What I'd really like to do is pick the whole city up, spray on a membrane, and drop it back down," he went on. I thought of Calvino's *Invisible Cities,* where such fantastical engineering schemes are the norm.

Mowry said he was intrigued by the possibility of finding some kind of resin that could be injected into the limestone. The resin would fill the holes, then set to form a seal. Or, he suggested, perhaps one day the city would require that builders, before constructing a house, lay a waterproof shield underneath it, the way a camper spreads a tarp under a tent. Or maybe some sort of clay could be pumped into the ground that would ooze out and fill the interstices.

"Will it hold?" Mowry said of the clay. "I doubt it. But these are things we're exploring." It was hard to tell how seriously he took any of these ideas; even if one of them turned out to be workable, the effort required to, in effect, caulk the entire island seemed staggering. At one point, Mowry declared, "If we can put a man on the moon, then we can figure out a way to keep Miami Beach dry." At another, he mused about the city's reverting to "what it came from," which was largely mangrove swamp: "I'm sure if we had poets, they'd be writing about the swallowing of Miami Beach by the sea."

We headed back toward Mowry's office around the time of maximum high tide. The elevated streets were still dry, but on the way to City Hall we came to an unreconstructed stretch of road that was flooding. Evidently, this situation had been anticipated, because two mobile pumps, the size and shape of ice-cream trucks, were parked near the quickly expanding pool. Neither was operating. After making a couple of phone calls, Mowry decided that he would try to switch them on himself. As he fiddled with the controls, I realized that we were standing not far from the drowned palm tree I'd seen on my first day in Miami Beach, and that it was once again underwater.

About a dozen miles due west of Miami, the land gives out, and what's left of the Everglades begins. The best way to get around in this part of Florida is by airboat, and on a gray morning I set out in one with a hydrologist named Christopher McVoy. We rented the boat from a concession run by members of the Miccosukee tribe, which, before the Europeans arrived, occupied large swaths of Georgia and Tennessee. The colonists hounded the Miccosukee ever farther south, until, eventually, they ended up with a few hundred mostly flooded square miles between Miami and Naples. On a fence in front of the dock, a sign read, BEWARE: WILD ALLIGATORS ARE DANGEROUS. DO NOT FEED OR TEASE. Our guide, Betty Osceola, handed out headsets to block the noise of the rotors, and we zipped off.

The Everglades is often referred to as a "river of grass," but it might just as accurately be described as a prairie of water. Where the airboats had made a track, the water was open, but mostly it was patchy—interrupted by clumps of sawgrass and an occasional tree island. We hadn't been out very long when it started to pour. As the boat sped into the rain, it felt as if we were driving through a sandstorm.

The same features that now make South Florida so vulnerable —its flatness, its high water table, its heavy rains—are the features that brought the Everglades into being. Before the drainage canals were dug, water flowed from Lake Okeechobee, about 70 miles north of Miami, to Florida Bay, about 40 miles to the south of the city, in one wide, slow-moving sheet. Now much of the water is diverted, and the water that does make it to the wetlands gets impounded, so the once-continuous "sheet flow" is no more. There's a comprehensive Everglades restoration plan, which goes by the acronym CERP, but this has got hung up on one political snag after another, and climate change adds yet one more obstacle. The Everglades is a freshwater ecosystem; already, at the southern margin of Everglades National Park, the water is becoming salty. The sawgrass is in retreat, and mangroves are moving in. In coming decades, there's likely to be more and more demand for the fresh water that remains. As McVoy put it, "You've got a big chunk of agriculture, a big chunk of people, and a big chunk of nature reserve all competing for the same resources."

The best that can be hoped for with the restoration project is that it will prolong the life of the wetland and, with that, of Miami's drinking-water system. But you can't get around geophysics. Send the ice sheets into "irreversible decline," as it seems increasingly likely we have done, and there's no going back. Eventually the Everglades, along with Shorecrest and Miami Beach and much of the rest of South Florida, will be inundated. And, if Hal Wanless is right, eventually isn't very far off.

To me, the gunmetal expanse of water and grass appeared utterly without markers, but Osceola, who could read the subtlest of ridges, knew exactly where we were at every moment. We stopped to have sandwiches on an island with enough dry land for a tiny farm, and stopped again at a research site that McVoy had set up in the muck. There was a box of electrical equipment on stilts, and a solar panel to provide power. McVoy dropped out of the boat to collect some samples in empty water-cooler bottles. The rain let up, and then started again.

KEA KRAUSE

What's Left Behind

FROM *The Believer*

IF YOU'RE FROM SEATTLE, like me, you learn early in life that Montana is spacious, touristy, and full of wayward relatives who knocked off the grid a long time ago. You know about Glacier and Yellowstone and the lax speed limits on the swaths of flat, endless highway beneath limitless skies. And of the few big towns in the state, you know sparse details: Helena is the capital, Missoula is a liberal stronghold, and in Butte a flooded copper mine—the nation's biggest body of toxic water, called the Berkeley Pit—functions as a town monument, a plaguing reminder of the price of industry, and, for some, a lab of curiosity. Montana is a weird, wide-open space—it's the 4th-largest state in the country, but 48th in population density; a place where you can still write personal checks for groceries, where bars feature attractions like live mermaids, and where Americans and mine waste alike are seemingly left alone to do whatever they want.

For years, as you approached Butte along I-90, all-you-can-eat-buffet-style billboards recommended the bizarre detour of the Berkeley Pit, marketing mine waste as historic pollution worth visiting. A massive hole filled with battery-acid-strength water, the signs suggested, isn't a far stretch from picnicking at a battleground or an old fort, retired sites from a different sort of war. Eventually, administrators realized that advertising the pit as a tourist attraction was damning to the town's reputation and took down the enticing signage, but visitors can still pay two dollars and, from a viewing stand, enjoy a recorded history of the town and the breathtaking vista of one of the greatest American copper-mining calamities of the 20th century.

Butte's history has all the heroic and romantic trappings of Wallace Stegner's nostalgic frontier saga *Angle of Repose*. After fortune-seekers panning for gold in Butte in the 1850s couldn't find any, the town was nearly left to return to nature. With only a handful of tacit laws keeping the peace, and without a mother lode, most men moved on. But miners working for one persevering entrepreneur named Marcus Daly, who had the copper version of the Midas touch, discovered a massive vein of the brown metal in 1882 and transformed Butte into the biggest copper-producing city in North America and, at one point, the entire world.

Upon my first visit I knew about the grandiose Butte lore and the pit, and I knew the word "perpetuity," which the Environmental Protection Agency appropriated while deeming the pit a Superfund site under the EPA's remediation program. The word was potent and suggested lifetimes: of scars, of people, of a pit—challenges that come without instructions. Everyone deals with their own disasters in perpetuity, and at the time I had my own: gray teeth and a crisscrossed lip from an accident years ago, a hastily instigated breakup, friendships lost in gulfs of my neglect. This was how the people of Butte would deal with the pit as well: perpetually, for a very long time.

The morning I met Joe Griffin, the state of Montana's Department of Environmental Quality representative, I got into his car without knowing where he was taking me. My Virgil in a dusty Subaru, Griffin led me on a twisting road away from Uptown Butte, through intermittent neighborhoods with boarded-up Craftsman bungalows, rusted-out cars, and the skeletal remains of bars and businesses, their irrelevant signs still dangling from chains with joints that creaked in the light summer breeze.

We came to a final stop in front of the fenced-in Bell Diamond mine, its gate guarded with a heavy chain. The air smelled of hot springs and hard-boiled eggs, and the elemental presence of sulfur clung to my hair and clothes like campfire smoke. Unbeknownst to me, the reason we were at this particular mine wasn't to observe the steel, mantis-shaped headframe that once lowered men and horses nearly a mile underground, nor to marvel at the ground itself, sparkling with feldspar and pyrite like a mirror-ball. Instead, we had come to take in the view of the Berkeley Pit, which, upon first take, surprised me with its similarity to a natural lake that might hold minnows, boats, and swimmers.

Panoramically, Butte doesn't make a whole lot of sense. What most people would consider a downtown area—a space with tall buildings, museums, bars, restaurants, and government complexes—in Butte is called "Uptown" because it's higher than everywhere else. Everywhere else is called "the Flats": a suburban valley populated with big-box stores and car dealerships. Uptown, the metropolitan center, is an art deco masterpiece, a six-square-block area with a cluster of nationally protected buildings that are handsome and important the way Wall Street's once were. But overshadowing its stately and peaceful vista was the treeless pit, looking diamond-cut with its precise edges forming a mile-long, half-mile-wide arrowhead filled with water so brown it looked thick.

The damage of the pit, in its absurd scale relative to the town, signals the historic pillaging of the land. The discovery of copper beneath Butte coincided with the development of the filament light bulb, a product whose mass production necessitated an abundance of copper wiring. Since Montana would not become a state for another decade, copper companies were left to pursue their own interests without regulation. Butte became an ant farm, with mining corporations ultimately digging out 10,000 miles' worth of tunnels under the town, a distance that could comfortably span from New York to Singapore.

Heap roasting was a common technology that used heat to convert sulfides in crushed rock into oxides that could then be smelted and refined into valuable ore. In the process, piles of rock, often the size of city blocks, were set on fire and allowed to burn for days. Smelting was just as noxious, producing excesses of smoke and another form of waste called slag, a muddy slush that would then get dumped from factories into nearby waterways. Silver Bow Creek, which runs through Butte, became a flowing mine-waste disposal site. Smoke from smelting engulfed the town; residents complained about not being able to see across streets. Cattle and other livestock died of arsenic poisoning. Trees ceased to grow.

In 1955 Anaconda Copper, the largest mining company in Butte, adopted the technique of open-pit mining, whereby land is terraced away to create a spiraling hole in the ground. The company proceeded to dig the Berkeley Pit, a hole big enough to accommodate the Eiffel Tower, near the center of Uptown Butte, displacing hundreds of residents and ruining the morale of the blue-collar mining community.

Some geological nuances contributed to the environmental problems developing in Butte. The ground beneath the town is an alluvial aquifer and consists of a watery porousness similar to the way fish-tank rocks sit loosely together. One of several abundant minerals in this soil is pyrite—fool's gold—which, when exposed to air and water, produces sulfuric acid. For years, companies spent significant time and money pumping groundwater out of mine shafts to accommodate tunneling miners, inadvertently also keeping Butte's pyrite dry. But when the oil company ARCO bought Anaconda in the late 1970s, copper was at its lowest price in years. To save money, ARCO decommissioned the nearly 100-year-old pumps. As the water rose, it reached the pyrite in the porous soil, producing sulfuric acid and mixing with the already existing pollutants. In 1983, once the tunnels flooded, the next thing to fill was the pit.

Catastrophe-wise, the pit falls in the middle of the spectrum. The Bingham Canyon copper mine, outside Salt Lake City, is nearly three times its size. Entire coal towns, like Centralia, Pennsylvania, have been evacuated because inextinguishable fires burn for years in mines beneath their streets. Though Griffin and I gazed out on a scene that was conclusively bad for the environment, I would learn that the pit struggles with a Goldilocks dilemma. While it isn't more disturbing than other sites, it remains the biggest body of contaminated water in the United States, and is essentially a walkable mile from Uptown—so close you could run errands to it on a lunch break. A sense of urgency also distinguishes the pit from those other sites: the contaminated water here is rising. If it reaches what the EPA has designated the "critical water level," which it is expected to do by 2023 without intervention, it will reverse its course and flow back into the water table. To prevent this, the plan put in place via Superfund is to pump and treat the pit water in perpetuity.

The Berkeley Pit is not likely to evacuate Butte. But to keep it safe and contained will drain money—millions of dollars—and attention, for whichever exists longer, the pit or us. All of southwest Montana is one big environmental liability in this sense: the region is infested with abandoned mines, an estimated 20,000 of them, each coming with its own hazardous idiosyncrasies and resting above the same watershed, which rushes out to the Pacific.

"If we all intend to keep going as a civilization, as a society, we still need copper. You're not going to stop progress. Nobody wants to give anything up, even environmentalists. It's a dilemma, really," Griffin mused. Even as we spoke, the Pebble Mine, in Bristol Bay, Alaska, one of the last unharmed watersheds in our country and the spawning ground of multiple endangered salmon breeds, was being proposed as a new copper venture.

"Obviously this isn't going anywhere. I mean, even if you thought, Let's take it somewhere, where the hell would you take it? So this will be here forever," Griffin said. The summer heat pushed down on us as the silent pit, an accident whose full resonance is still unknown, stretched far off to the edges of Butte.

I wanted to make sure I had heard Griffin right, that he had actually used the word "forever" in our conversation, so I consulted a hydrogeologist. Nick Tucci, who worked on the Berkeley Pit for nearly a decade, clarified some details. Tucci moved to Butte in 2003 to get a master's degree in geoscience from Montana Tech, an outpost of the University of Montana and one of the nation's preeminent mining colleges. For Tucci, Butte, with its flooded pit mine, was the perfect place to study. Tucci is articulate and sincere, and bends your ear about complicated subjects in a way that makes you feel like you are discussing something as accessible as the weather.

"There's no proven technology out there that we have right now to stop the water from infiltrating the pit," he told me. "I think whatever technology you use that exists right now, you are going to be pumping and treating forever. Right now, we just don't have the ability to clean up the Berkeley Pit."

But that hasn't been for lack of trying. In his time working at the Montana Bureau of Mines and Geology, Tucci had given out samples of pit water to varying types of scientists (some self-declared) eager to try to solve the problem. Tucci witnessed ideas that varied from snake oil to brilliant. "There was a guy who brought in crystals and thought he could arrange them in such a way that it wouldn't take the contaminants out of the water but would rearrange the molecules to the extent where they were no longer toxic. He was convinced it would work and he was going to show us by drinking it."

Other ideas involved evaporating the pit water with mirrors or sprinkler heads spraying the water over fields where it would evaporate, a method that has worked with other pit lakes (though there would still be the conundrum of water quality to address). One idea that Tucci liked, but wasn't sure was feasible because of its cost, was to fill the pit with mine tailings, a type of mine waste that's like dirt. Groundwater would still have to be pumped in perpetuity, but at least the surface area of the pit could be used for something.

Even without the end goal of remediation, the pit has served as a laboratory for diverse experimentation. Andrea and Don Stierle, organic chemists and old friends of both Griffin's and Tucci's, grew interested in the pit water when they moved to Butte from San Diego. While Don taught at Montana Tech, Andrea, who had originally planned to be a marine natural products chemist, decided to explore Butte's largest body of water, toxic or not. It's not uncommon for scientists to pursue bacteria growth in mine waste —anything thriving in a toxic spot could hold the key to its own cleanup—but no one had really looked into acidic mine water as something that could support life. To Andrea, life in the pit didn't seem too far-fetched. It posed a risky but thrilling challenge. She doesn't dumb things down, so here's how she explained it: "Nobody had looked in toxic waste for a bacterium or a fungus that could produce secondary metabolites that had biological activity that could be helpful. It seemed like a great idea!" It did.

The Stierles weren't initially able to isolate any sort of living organism in the water—until a flock of snow geese spent a perilous night floating on the pit's surface, resting their wings after flying directly into a storm. A black yeast that hadn't been there before started to grow in the water. Andrea sent the culture to a lab for identification and learned something unexpected: the yeast was associated with goose rectums. For years, scientists had been unsuccessfully throwing organic matter, like hay or horse manure, into the pit to see if it would yield new life. The night of the storm, when the party of geese realized that what they were treading wasn't exactly water, they fled the toxic pond. When birds take flight they evacuate their bowels to lighten their load and ease the process of takeoff. In this instance, an entire flock evacuated simultaneously and filled the pit with biological matter. Goose poop had made the pit come alive.

Prior to the discovery of the yeast, Andrea Stierle had already had many successes in her career. While in San Diego, she held a postdoctoral position at the Scripps Institution of Oceanography, and in the 1990s she discovered a fungus in the bark of the Pacific yew tree that is used in the formulation of Taxol, a cancer-fighting drug. Her focus was trained on the minuscule, and her process has always been to reduce her subjects to microscopic proportions and see what biological activity she could find.

In the pit, the Stierles had set out to find biological life with medicinal properties, similar to Taxol's cancer-fighting capabilities. In fact, the yeast was showing promise as having an effect on two types of ovarian cancer cell lines. But the yeast had another exciting attribute: it pulled metals out of Berkeley Pit water.

Acid-generating rock in the ground beneath Butte creates metal finds—or clusters—that clog filters and amass in wells. In the case of the pit, the finds are high in iron, which is responsible for the water's blood-red color. And as Andrea learned through her studies, the slime liked to eat, or sorb, iron. "This is one of the things this little yeast does exquisitely. It takes these finds, and if you add a drop of the yeast when we grow it in liquid solution and add it to a big flask of pit water with the finds, this yeast will take all the finds, drop them out of solution, and just form this little blob," she explained. "It's amazing. It quickly adsorbs up to 87 percent of a lot of the metals that are present, and that happens within five to ten seconds. It's almost magical."

In her lab, Andrea held a pipette with the yeast cocked over a beaker filled with Berkeley Pit water and dripped its viscous contents into the beaker. "Let's go, baby!" she encouraged the yeast. She swirled the beaker like a lowball glass of whiskey, the yeast twining in dark, wispy tentacles. In the midst of the swishing, the water's cloudiness diminished and a black, marble-sized ball formed in the beaker's center. The yeast had sorbed the metals that once polluted the water, which now did look clean enough to drink. Her slime was a glimpse of refreshing innovation, a repurposing of disaster.

*

Since grant money for people like the Stierles is scarce and there's a dearth of sustaining jobs in Butte, I wanted to know who was young and brave enough to commit the burgeoning parts of their career to a town that is, for all intents and purposes, extremely economically depressed. Griffin pointed me in the direction of Julia Crain, the special projects planner for the Butte–Silver Bow consolidated city-county government, who is also involved in the Superfund program. Crain is a third-generation Butte resident —her grandfather helped build one of the town's first railroads —and holds a graduate degree in urban and regional planning from Portland State University.

Crain is ambitious and devoted to Butte, and Griffin, Tucci, and Stierle all agree that if Butte has one good thing going for it, it's Julia. She inexhaustibly writes grants for things that residents don't even realize they deserve. In its risk, hard-rock mining once represented the pinnacle of manhood, with miners working hard in the wretched conditions underground and living hard in the bars and brothels above. Butte is still a tough town, and this attitude can stand in the way of progressive change. But Crain is undeterred, and has been awarded millions of dollars to build recreational trails through Butte's public greenspaces and plant trees along Uptown streets, amenities the town didn't know it missed until it had them. Butte's been "taken hostage by its perception of itself," Crain says, in regard to its proud reputation as an overbold frontier town—which can cause her work, along with elements of the cleanup, to be met with occasional hostility.

But that's not necessarily a bad thing. "We know that dialogue here is really healthy and that people are really engaged, because every issue has contention surrounding it, and it's because people are holding fast to something they love. I don't think they're saying no to ushering in a new era; I think they want to be confident there's someone there to really carry it forward into the forever." Change needs gentle coaxing in Butte, but Crain knows that blooming late is better than never blooming at all. "We are playing catch-up. We had to spend thirty years cleaning up a bunch of contamination and figuring out how to protect the people that live here so they could stay. So it's not as though we aren't progressing; it's just that we had to take a different approach to [progress] because of the situation we found ourselves in."

Despite opportunity existing all around in subtle forms, no one was moving to Butte or staying in Butte for it. Instead, one might stay for the town's sense of everlasting potential and the belief systems built around it—just like in the days when Montana was still a frontier. "I think everybody here has their own romances, and maybe some people don't have the words to express it, but I know that Andrea Stierle was completely enraptured by the Berkeley Pit, and her research is the result of that," Crain explained. "You have to be capable of seeing something more, and that's how people can get through here. I'm not saying it's that hard; I'm just saying it's a diamond in the rough."

Thirty-five hundred feet above Butte, along one of the Continental Divide's arched ridges, stands a surprising, 90-foot statue called Our Lady of the Rockies. Our Lady took six years to construct, from the initial plans in 1979 to the final portion, her head, which was airlifted by helicopter to the top of the mountain in the winter of 1985. She looms protectively over Butte, arms outstretched in a come-gimme-a-hug pose. She is impossible to miss from the streets of Butte. During one of the city's economic downturns, miners designed and welded her out of donated steel to serve as a symbol of workers everywhere. She is as strange and Herculean as everything else in Butte: the patron saint of toughness. As the third-tallest statue in the United States, and painted a scorching white, Our Lady is Butte's very own Christ the Redeemer. And you can visit her, twice a day, by way of a shuttle bus departing from the Butte Plaza Mall.

I took the morning tour to Our Lady to avoid the hot July sun. The narrow, unpaved road, the retiree bus driver, and the bus that should've been retired were a nerve-wracking combination during the 45 minutes we spent laboring up the mountain. At the top, I scurried around the base of Our Lady, taking pictures that never managed to get her full figure in the frame. Around me, kids kicked at bushes and people put quarters into mounted binoculars, but the majority of the tourists kneeled and touched Our Lady, seemingly in prayer. The few people I spoke to on the bus weren't from Butte, or even from Montana, but were on vacation, and, as people who were either still working or had worked blue-collar jobs, had a reason to be on that bus: they were paying homage to a town whose culture revolved around self-reliance and whose entire workforce had operated around hard, hazardous, thankless work for generations.

Without realizing it, I had begun to fancy myself a pilgrim, too. Being engulfed in the wilds of national parks still fills me with the awe I first experienced out on Washington's Olympic Peninsula, where I spent summers growing up. There, the land is safe, and nature has the right of way. And it was in a national park, in my late 20s, that I realized that some of my favorite sensory perceptions, like the bitter smell of ferns after rain and the sounds of creatures scurrying through dried leaves, exist largely in places that are now protected by the government.

The lines between what is nature and what is natural have become blurred in my lifetime. Other mammals also consume the land: in Butte, enterprising beavers have depleted floodplains all around the valley with their dams. In getting what you want as a species, it sometimes seems impossible not to leave some sort of mark, but it is the remnants that have become my main concern. Since the damage has already been done, I want to know what we are doing with the damage—how we are transforming our destruction into creation.

During one of our conversations, Tucci, the hydrogeologist, said something to me that I had been thinking all along but had been afraid to say out loud, which was that the pit was beautiful. At first I had wanted to say that it was hideous, sinister even, but the pit's engineered tiers, industrialized terra-cotta complexion, and crimson water have a hard-won refinement, like western art scenes of dusty cavalcades and buffalo runs. "I think the research value of the Berkeley Pit is not quantifiable and the lessons that can be learned from the Berkeley Pit are not quantifiable," Tucci said. "People are going to come here, look at the Berkeley Pit, and know what we are capable of, and people will be cautious, hopefully."

Though on the morning of my visit to Our Lady I felt as though I didn't deserve a seat on the bus, by midafternoon I had decided the miners who built her would want me to believe she belonged to me, too, and that I was welcome to join my fellow tourists in praying at the folds of her steel robes if I liked. Even when it doesn't seem like it, there is a lot of connectivity in towns like Butte that masquerade as the edge of the universe but are really its center. Had I known at the time that we were all there because embracing our scars as they amass is difficult, and loving hard-to-love things is alienating, maybe I would've rested my forehead against her cool, steel siding, too.

ROBERT KUNZIG

The Will to Change

FROM *National Geographic*

HAMBURG KNEW THE BOMBS were coming, and so the prisoners of war and forced laborers had just half a year to build the giant flak bunker. By July 1943 it was finished. A windowless cube of reinforced concrete, with seven-foot-thick walls and an even thicker roof, it towered like a medieval castle above a park near the Elbe River. The guns protruding from its four turrets would sweep Allied bombers from the sky, the Nazis promised, while tens of thousands of citizens sheltered safely behind its impenetrable walls.

Coming in at night from the North Sea just weeks after the bunker was finished, British bombers steered for the spire of St. Nikolai in the center of the city. They dropped clouds of metallic foil strips to throw off German radar and flak gunners. Targeting crowded residential neighborhoods, the bombers ignited an unquenchable firestorm that destroyed half of Hamburg and killed more than 34,000 people. Towering walls of fire created winds so strong that people were blown into the flames. Church bells clanged furiously.

The spire of St. Nikolai, which somehow survived, stands today as a *Mahnmal*—a memorial reminding Germany of the hell brought by the Nazis. The flak bunker is another *Mahnmal.* But now it has a new meaning: it has been transformed from a powerful reminder of Germany's shameful past into a hopeful vision for its future.

In the central space of the bunker, where people once cowered through the firestorm, a six-story, 528,000-gallon hot-water tank delivers heat and hot water to some 800 homes in the neighborhood. The water is warmed by burning gas from sewage treatment, by waste heat from a nearby factory, and by solar panels that now cover the roof of the bunker, supported by struts angling from the old gun turrets. The bunker also converts sunlight into electricity; a scaffolding of photovoltaic (PV) panels on its south façade feeds enough juice into the grid to supply a thousand homes. On the north parapet, from which the flak gunners once watched flames rising from the city center, an outdoor café offers a view of the changed skyline. It's dotted with 17 wind turbines now.

Germany is pioneering an epochal transformation it calls the *Energiewende*—an energy revolution that scientists say all nations must one day complete if a climate disaster is to be averted. Among large industrial nations, Germany is a leader. Last year about 27 percent of its electricity came from renewable sources such as wind and solar power, three times what it got a decade ago and more than twice what the United States gets today. The change accelerated after the 2011 meltdown at Japan's Fukushima nuclear power plant, which led Chancellor Angela Merkel to declare that Germany would shut all 17 of its own reactors by 2022. Nine have been switched off so far, and renewables have more than picked up the slack.

What makes Germany so important to the world, however, is the question of whether it can lead the retreat from fossil fuels. By later this century, scientists say, planet-warming carbon emissions must fall to virtually zero. Germany, the world's fourth-largest economy, has promised some of the most aggressive emission cuts—by 2020 a 40 percent cut from 1990 levels, and by 2050 at least 80 percent.

The fate of those promises hangs in the balance right now. The German revolution has come from the grassroots: individual citizens and energy *Genossenschaften*—local citizens' associations—have made half the investment in renewables. But conventional utilities, which didn't see the revolution coming, are pressuring Merkel's government to slow things down. The country still gets far more electricity from coal than from renewables. And the Energiewende has an even longer way to go in the transportation and heating sectors, which together emit more carbon dioxide (CO_2) than power plants.

German politicians sometimes compare the Energiewende to

the Apollo moon landing. But that feat took less than a decade, and most Americans just watched it on TV. The Energiewende will take much longer and will involve every single German—more than 1.5 million of them, nearly 2 percent of the population, are selling electricity to the grid right now. "It's a project for a generation; it's going to take till 2040 or 2050, and it's hard," said Gerd Rosenkranz, a former journalist at *Der Spiegel* who's now an analyst at Agora Energiewende, a Berlin think tank. "It's making electricity more expensive for individual consumers. And still, if you ask people in a poll, Do you want the Energiewende? then 90 percent say yes."

Why? I wondered as I traveled in Germany last spring. Why is the energy future happening here, in a country that was a bombed-out wasteland 70 years ago? And could it happen everywhere?

The Germans have an origin myth: it says they came from the dark and impenetrable heart of the forest. It dates back to the Roman historian Tacitus, who wrote about the Teutonic hordes who massacred Roman legions, and it was embellished by German romantics in the 19th century. Through the upheavals of the 20th century, according to ethnographer Albrecht Lehmann, the myth remained a stable source of German identity. The forest became the place where Germans go to restore their souls—a habit that predisposed them to care about the environment.

So in the late 1970s, when fossil fuel emissions were blamed for killing German forests with acid rain, the outrage was nationwide. The oil embargo of 1973 had already made Germans, who have very little oil and gas of their own, think about energy. The threat of *Waldsterben,* or forest death, made them think harder.

Government and utilities were pushing nuclear power—but many Germans were pushing back. This was new for them. In the decades after World War II, with a ruined country to rebuild, there had been little appetite for questioning authority or the past. But by the 1970s the rebuilding was complete, and a new generation was beginning to question the one that had started and lost the war. "There's a certain rebelliousness that's a result of the Second World War," a 50-something man named Josef Pesch told me. "You don't blindly accept authority."

Pesch was sitting in a mountaintop restaurant in the Black Forest outside Freiburg. In a snowy clearing just uphill stood two

320-foot-tall wind turbines funded by 521 citizen investors recruited by Pesch—but we weren't talking about the turbines yet. With an engineer named Dieter Seifried, we were talking about the nuclear reactor that never got built, near the village of Wyhl, 20 miles away on the Rhine River.

The state government had insisted that the reactor had to be built or the lights would go out in Freiburg. But beginning in 1975, local farmers and students occupied the site. In protests that lasted nearly a decade, they forced the government to abandon its plans. It was the first time a nuclear reactor had been stopped in Germany.

The lights didn't go out, and Freiburg became a solar city. Its branch of the Fraunhofer Institute is a world leader in solar research. Its Solar Settlement, designed by local architect Rolf Disch, who'd been active in the Wyhl protests, includes 50 houses that all produce more energy than they consume. "Wyhl was the starting point," Seifried said. In 1980 an institute that Seifried cofounded published a study called *Energiewende*—giving a name to a movement that hadn't even been born yet.

It wasn't born of a single fight. But opposition to nuclear power, at a time when few people were talking about climate change, was clearly a decisive factor. I had come to Germany thinking the Germans were foolish to abandon a carbon-free energy source that, until Fukushima, produced a quarter of their electricity. I came away thinking there would have been no Energiewende at all without antinuclear sentiment—the fear of meltdown is a much more powerful and immediate motive than the fear of slowly rising temperatures and seas.

All over Germany I heard the same story. From Disch, sitting in his own cylindrical house, which rotates to follow the sun like a sunflower. From Rosenkranz in Berlin, who back in 1980 left physics graduate school for months to occupy the site of a proposed nuclear-waste repository. From Luise Neumann-Cosel, who occupied the same site two decades later—and who is now leading a citizens' initiative to buy the Berlin electric grid. And from Wendelin Einsiedler, a Bavarian dairy farmer who has helped transform his village into a green dynamo.

All of them said Germany had to get off nuclear power and fossil fuels at the same time. "You can't drive out the devil with Beelzebub," explained Hans-Josef Fell, a prominent Green Party politician. “Both have to go.” At the University of Applied Sciences in Berlin, energy researcher Volker Quaschning put it this way: “Nuclear power affects me personally. Climate change affects my kids. That’s the difference.”

If you ask why antinuclear sentiment has been so much more consequential in Germany than, say, across the Rhine in France, which still gets 75 percent of its electricity from nukes, you end up back at the war. It left Germany a divided country, the front along which two nuclear superpowers faced off. Demonstrators in the 1970s and ’80s were protesting not just nuclear reactors but plans to deploy American nuclear missiles in West Germany. The two didn’t seem separable. When the German Green Party was founded in 1980, pacifism and opposition to nuclear power were both central tenets.

In 1983 the first Green representatives made it into the Bundestag, the national parliament, and started injecting green ideas into the political mainstream. When the Soviet reactor at Chernobyl exploded in 1986, the left-leaning Social Democrats (SPD), one of Germany’s two major parties, was converted to the antinuclear cause. Even though Chernobyl was hundreds of miles away, its radioactive cloud passed over Germany, and parents were urged to keep their children inside. It’s still not always safe to eat mushrooms or wild boar from the Black Forest, Pesch said. Chernobyl was a watershed.

But it took Fukushima, 25 years later, to convince Merkel and her Christian Democratic Union (CDU) that all nuclear reactors should be switched off by 2022. By then the boom of renewable energy was in full swing. And a law that Hans-Josef Fell had helped create back in 2000 was the main reason.

Fell’s house in Hammelburg, the town in northern Bavaria where he was born and raised, is easy to spot among all the pale postwar stucco: it’s the one built of dark larch wood, with a grass roof. On the south side, facing the backyard, the grass is partially covered by photovoltaic and solar hot-water panels. When there’s not enough sun to produce electricity or heat, a cogenerator in the basement burns sunflower or rapeseed oil to produce both. On the March morning when I visited, the wood interior of the house was bathed in sunlight and warmth from the conservatory. In a few weeks, Fell said, wildflowers would be blooming on the roof.

A tall man in jeans and Birkenstocks, with a bald, egg-shaped head and a fringe of gray beard, Fell has moments of sounding like a preacher—but he's no green ascetic. A shed in his backyard, next to the swimming pond, houses a sauna, powered by the same green electricity that powers his house and his car. "The environmental movement's biggest mistake has been to say, 'Do less. Tighten your belts. Consume less,'" Fell said. "People associate that with a lower quality of life. 'Do things differently, with cheap, renewable electricity'—that's the message."

From Fell's garden, on a clear day, you used to be able to see the white steam plumes of the nuclear reactor at Grafenrheinfeld. His father, the conservative mayor of Hammelburg, supported nuclear power and the local military base. Young Fell demonstrated at Grafenrheinfeld and went to court to refuse military service. Years later, after his father had retired, Fell was elected to the Hammelburg city council.

It was 1990, the year Germany was officially reunified—and while the country was preoccupied with that monumental task, a bill boosting the Energiewende made its way through the Bundestag without much public notice. Just two pages long, it enshrined a crucial principle: producers of renewable electricity had the right to feed into the grid, and utilities had to pay them a "feed-in tariff." Wind turbines began to sprout in the windy north.

But Fell, who was installing PV panels on his roof in Hammelburg, realized that the new law would never lead to a countrywide boom: it paid people to produce energy, but not enough. In 1993 he got the city council to pass an ordinance obliging the municipal utility to guarantee any renewable-energy producer a price that more than covered costs. Fell promptly organized an association of local investors to build a 15-kilowatt solar power plant —tiny by today's standards, but the association was one of the first of its kind. Now there are hundreds in Germany.

In 1998 Fell rode a Green wave and his success in Hammelburg into the Bundestag. The Greens formed a governing coalition with the SPD. Fell teamed up with Hermann Scheer, a prominent SPD advocate of solar energy, to craft a law that in 2000 took the Hammelburg experiment nationwide and has since been imitated around the world. Its feed-in tariffs were guaranteed for 20 years, and they paid well.

"My basic principle," Fell said, "was the payment had to be so high that investors could make a profit. We live in a market economy, after all. It's logical."

Fell was about the only German I met who claimed not to have been surprised at the boom his logic unleashed. "That it would be possible to this extent—I didn't believe that then," said dairy farmer Wendelin Einsiedler. Outside his sunroom, which overlooks the Alps, nine wind turbines turned lazily on the ridge behind the cow pen. The smell of manure drifted in. Einsiedler had started his personal Energiewende in the 1990s with a single turbine and a methane-producing manure fermenter. He and his brother Ignaz, also a dairy farmer, burned the methane in a 28-kilowatt cogenerator, generating heat and electricity for their farms. "There was no question of making money," Einsiedler said. "It was idealism."

But after the renewable-energy law took effect in 2000, the Einsiedlers expanded. Today they have five fermenters, which process corn silage as well as manure from eight dairy farms, and they pipe the resulting biogas three miles to the village of Wildpoldsried. There it's burned in cogenerators to heat all the public buildings, an industrial park, and 130 homes. "It's a wonderful principle, and it saves an unbelievable amount of CO_2," said Mayor Arno Zengerle.

The biogas, the solar panels that cover many roofs, and especially the wind turbines allow Wildpoldsried to produce nearly five times as much electricity as it consumes. Einsiedler manages the turbines, and he's had little trouble recruiting investors. Thirty people invested in the first one; 94 jumped on the next. "These are their wind turbines," Einsiedler said. Wind turbines are a dramatic and sometimes controversial addition to the German landscape—"asparagification," opponents call it—but when people have a financial stake in the asparagus, Einsiedler said, their attitude changes.

It wasn't hard to persuade farmers and homeowners to put solar panels on their roofs; the feed-in tariff, which paid them 50 cents a kilowatt-hour when it started in 2000, was a good deal. At the peak of the boom, in 2012, 7.6 gigawatts of PV panels were installed in Germany in a single year—the equivalent, when the sun is shining, of seven nuclear plants. A German solar-panel industry blossomed, until it was undercut by lower-cost manufacturers in China—which took the boom worldwide.

Fell's law, then, helped drive down the cost of solar and wind, making them competitive in many regions with fossil fuels. One sign of that: Germany's tariff for large new solar facilities has fallen from 50 euro cents a kilowatt-hour to less than 10. "We've created a completely new situation in 15 years—that's the huge success of the renewable-energy law," Fell said.

Germans paid for this success not through taxes but through a renewable-energy surcharge on their electricity bills. This year the surcharge is 6.17 euro cents per kilowatt-hour, which for the average customer amounts to about 18 euros a month—a hardship for some, Rosenkranz told me, but not for the average German worker. The German economy as a whole devotes about as much of its gross national product to electricity as it did in 1991.

In the 2013 elections Fell lost his seat in the Bundestag, a victim of internal Green Party politics. He's back in Hammelburg now, but he doesn't have to look at the steam plumes from Grafenrheinfeld: last June the reactor became the latest to be switched off. No one, not even the industry, thinks nuclear is coming back in Germany. Coal is another story.

Germany got 44 percent of its electricity from coal last year—18 percent from hard coal, which is mostly imported, and about 26 percent from lignite, or brown coal. The use of hard coal has declined substantially over the past two decades, but not the use of lignite. That's a major reason Germany isn't on track to meet its own greenhouse gas emissions target for 2020.

Germany is the world's leading producer of lignite. It emits even more CO_2 than hard coal, but it's the cheapest fossil fuel —cheaper than hard coal, which is cheaper than natural gas. Ideally, to reduce emissions, Germany should replace lignite with gas. But as renewables have flooded the grid, something else has happened: on the wholesale market where contracts to deliver electricity are bought and sold, the price of electricity has plummeted, such that gas-fired power plants and sometimes even plants burning hard coal are priced out of the market. Old lignite-fired power plants are rattling along at full steam, 24-7, while modern gas-fired plants with half the emissions are standing idle.

"Of course we have to find a track to get rid of our coal—it's very obvious," said Jochen Flasbarth, state secretary in the environment ministry. "But it's quite difficult. We are not a very resource-rich country, and the one resource we have is lignite."

Curtailing its use is made harder by the fact that Germany's big utilities have been losing money lately—because of the Energiewende, they say; because of their failure to adapt to the Energiewende, say their critics. E.ON, the largest utility, which owns Grafenrheinfeld and many other plants, declared a loss of more than 3 billion euros last year.

"The utilities in Germany had one strategy," Flasbarth said, "and that was to defend their track—nuclear plus fossil. They didn't have a strategy B." Having missed the Energiewende train as it left the station, they're now chasing it. E.ON is splitting into two companies, one devoted to coal, gas, and nuclear, the other to renewables. The CEO, once a critic of the Energiewende, is going with the renewables.

Vattenfall, a Swedish state-owned company that's another one of Germany's four big utilities, is attempting a similar evolution. "We're a role model for the Energiewende," spokesperson Lutz Wiese said cheerfully as he greeted me at Welzow-Süd—an open-pit mine on the Polish border that produces 22 million tons of lignite a year. In a trench that covers 11 square miles and is more than 300 feet deep, 13 gargantuan digging machines work in synchrony—moving the trench through the landscape, exposing and removing the lignite seam, and dumping the overburden behind them so the land can be replanted. In one recultivated area there's a small experimental vineyard. On the same rebuilt hill stands a memorial to Wolkenberg, a village consumed by the mine in the 1990s. Boulders mark the spots where the church and other buildings once stood.

It was a gorgeous spring day; from Wolkenberg, the only cloud we could see was the lazily billowing steam plume from the 1.6-gigawatt power plant at Schwarze Pumpe, which burns most of the lignite mined at Welzow-Süd. In a conference room, Olaf Adermann, asset manager for Vattenfall's lignite operations, explained that Vattenfall and other utilities had never expected renewables to take off so fast. Even with the looming shutdown of more nuclear reactors, Germany has too much generating capacity.

"We have to face some kind of a market cleaning," Adermann said. But lignite shouldn't be the one to go, he insisted: it's the "reliable and flexible partner" when the sun isn't shining or the wind isn't blowing. Adermann, who's from the region and worked for its lignite mines before they belonged to Vattenfall, sees them continuing to 2050—and maybe beyond.

Vattenfall, however, plans to sell its lignite business, if it can find a buyer, so it can focus on renewables. It's investing billions of euros in two new offshore wind parks in the North Sea—because there's more wind offshore than on and because a large corporation needs a large project to pay its overhead. "We can't do onshore in Germany," Wiese said. "It's too small."

Vattenfall isn't alone: the renewables boom has moved into the North and Baltic Seas and, increasingly, into the hands of the utilities. Merkel's government has encouraged the shift, capping construction of solar and onshore wind and changing the rules in ways that shut out citizens' associations. Last year the amount of new solar fell to around 1.9 gigawatts, a quarter of the 2012 peak. Critics say the government is helping big utilities at the expense of the citizens' movement that launched the Energiewende.

At the end of April, Vattenfall formally inaugurated its first German North Sea wind park, an 80-turbine project called DanTysk that lies some 50 miles offshore. The ceremony in a Hamburg ballroom was a happy occasion for the city of Munich too. Its municipal utility, Stadtwerke München, owns 49 percent of the project. As a result Munich now produces enough renewable electricity to supply its households, subway, and tram lines. By 2025 it plans to meet all of its demand with renewables.

In part because it has retained a lot of heavy industry, Germany has some of the highest per capita carbon emissions in western Europe. (They're a bit more than half of U.S. emissions.) Its goal for 2020 is to cut them by 40 percent from 1990 levels. As of last year, it had achieved 27 percent. The European carbon-trading system, in which governments issue tradable emissions permits to polluters, hasn't been much help so far. There are too many permits in circulation, and they're so cheap that industry has little incentive to cut emissions.

Though Germany isn't on track to meet its own goal for 2020, it's ahead of the European Union's schedule. It could have left things there—and many in Merkel's CDU wanted her to do just that. Instead, she and Economic Affairs Minister Sigmar Gabriel, head of the SPD, reaffirmed their 40 percent commitment last fall.

They haven't proved they can meet it, however. Last spring Gabriel proposed a special emissions levy on old, inefficient coal plants; he soon had 15,000 miners and power plant workers, encouraged by their employers, demonstrating outside his ministry. In July the government backed down. Instead of taxing the utilities, it said it would pay them to shut down a few coal plants—achieving only half the planned emissions savings. For the Energiewende to succeed, Germany will have to do much more.

It will have to get off gasoline and diesel too. The transportation sector produces about 17 percent of Germany's emissions. Like the utilities, its famous carmakers—Mercedes-Benz, BMW, Volkswagen, and Audi—were late to the Energiewende. But today they're offering more than two dozen models of electric cars. The government's goal is to have a million electric cars on the road by 2020; so far there are about 40,000. The basic problem is that the cars are still too expensive for most Germans, and the government hasn't offered serious incentives to buy them—it hasn't done for transportation what Fell's law did for electricity.

Much the same is true of buildings, whose heating systems emit 30 percent of Germany's greenhouse gases. Rolf Disch in Freiburg is one of many architects who have built houses and buildings that consume almost no net energy or produce a surplus. But Germany is not putting up many new buildings. "The strategy has always been to modernize old buildings in such a way that they use almost no energy and cover what they do use with renewables," said Matthias Sandrock, a researcher at the Hamburg Institute. "That's the strategy, but it's not working. A lot is being done, but not enough."

All over Germany, old buildings are being wrapped in six inches of foam insulation and refitted with modern windows. Low-interest loans from the bank that helped rebuild the war-torn west with Marshall Plan funds pay for many projects. Just 1 percent of the stock is being renovated every year, though. For all buildings to be nearly climate-neutral by 2050—the official goal—the rate would need to double at least. Once, Sandrock said, the government floated the idea of requiring homeowners to renovate. The public outcry shot that trial balloon down.

*

"After Fukushima, for a short time there was *Aufbruchstimmung*—for about half a year there was a real euphoria," said Gerd Rosenkranz. *Aufbruchstimmung* means something like "the joy of departure"; it's what a German feels when he's setting out on a long hike, say, in the company of friends. With all the parties in Germany in agreement, Rosenkranz said, the Energiewende felt like that. But the feeling hasn't lasted. Economic interests are clashing now. Some Germans say it might take another catastrophe like Fukushima to catalyze a fresh burst of progress. "The mood is bad," Rosenkranz said.

But here's the thing about the Germans: they knew the Energiewende was never going to be a walk in the forest, and yet they set out on it. What can we learn from them? We can't transplant their desire to reject nuclear power. We can't appropriate their experience of two great nation-changing projects—rebuilding their country when it seemed impossible, 70 years ago, and reunifying their country when it seemed forever divided, 25 years ago. But we can be inspired to think that the Energiewende might be possible for other countries too.

In a recent essay William Nordhaus, a Yale economist who has spent decades studying the problem of addressing climate change, identified what he considers its essence: free riders. Because it's a global problem, and doing something is costly, every country has an incentive to do nothing and hope that others will act. While most countries have been free riders, Germany has behaved differently: it has ridden out ahead. And in so doing, it has made the journey easier for the rest of us.

AMY LEACH

The Modern Moose

FROM *Ecotone*

RECENTLY, SOME MODERN ANIMALS have been reconsidering their attachment to the Earth. She was a lot of fun in her honeyed youth, but she's getting sick and seedy now, temperamental and pockmarked and tired. They find her decline kind of depressing, kind of repulsive. Before she is totally moribund they are looking around for other options, redirecting their attentions to places like the Kuiper Belt and the Large Magellanic Cloud, which contain untapped riches of rare materials and are unaffected by banana blight. In fact, such moderns have a hoof or two in heaven already: if they say yes to the Earth it's an equivocal yes, easy to disavow as soon as they can blast off to a better where.

But as ravishing as the Cloud/Belt may be, not all modern animals seem so eager to leave the Earth. Some are still entangled with seas and trees and Russian tundra; others seem entirely indifferent to the idea of space settlement. Take the modern moose, for example. Has he sent scouts to the moon? Has he shown any interest in starships? Has he ever practiced grooming himself in an antigravity gyroscopic device? No, no, and egads, no. Nevertheless the moose is as modern as Mugellini and should be coequally respected. In his survey of everything, *Modern History; or, The Present State of All Nations,* Thomas Salmon duly notes the concerns of the modern moose: he likes to chew on young shrubs, "but mostly, and with greatest delight, on water-plants, especially a sort of wild Colts-foot and Lilly that abound in our ponds, and by the sides of the rivers, and for which the Moose will wade far and deep."

With the magnitude of his antlers, it's not like the moose

could ever be flighty anyway. The first time bone starts coming out of his brow, the young moose might think it will turn into something moderate and flattering, something like a pillbox hat. Maybe his horns will be trinket horns, party horns, flirty horns like the giraffe's, or sleek Armani antlers like the pronghorn's. But the bumps grow into spikes, and the spikes spread and branch and keep growing, past trinket, past flirty, past flattering, and far past moderate. (Moderate horns are for moderate species; moderate species get very excited about moderate horns.) Finally they grow past preposterous: on his head the moose carries branching antlers so absurdly heavy he mustn't lower his head down to the ground, for fear he'll never raise it up again.

Seventy pounds of antler seems like an affirmation somehow, an exaggerated weight implacably attaching the moose to the Earth, saying, "Yes without a question, yes with all my heart. Yes if it pitches me face-first into the mud, yes if the rest of me withers, yes if my yes gets splintered, or broken, or deformed: yes and yes and yes again." When you have wings your hope can be elsewhere —up the hill, up the sky. But wings of solid bone scorn the idea of flight: your hope must be here.

Of course the moose did not choose his yes any more than Respighi chose his. It is his hap to be born—to come out of the mama onto the Earth—his hap to be rejected by the mama once a littler moose comes out, his hap to bear a staggering affirmation on his brow. A real yes weighs you down, like a woe, like those lunatic vows of lovers, to be kept even if one finds a better who. *I,* Alces alces, *take you Earth to be my planet, to have and to hold, for better or for worse, for richer, for poorer, in sickness and in health, to love and to cherish; from this day forward until death do us part.* Yes to the pond where the water-plants thrive, yes to the pond where the water-plants fail, yes to the pock where the pond used to be. Yes to you healthy, yes to you sick, yes to you blooming, and yes to you stricken. Though you have seen better days, though you no longer delight me with Colts-foot and Lillies, yes to the Earth, my Earth, for I do not hope to find a better where.

APOORVA MANDAVILLI

The Lost Girls

FROM *Spectrum*

IT TOOK 10 years, 14 psychiatrists, 17 medications, and 9 diagnoses before someone finally realized that what Maya has is autism. Maya loves numbers, and with her impeccable memory, she can rattle off these stats: that the very first psychiatrist she saw later lost his right to practice because he slept with his patients. That psychiatrist number 12 met with her for all of seven minutes and sent her out with no answers. That during her second year at Cambridge University in the United Kingdom, industrial doses of the antipsychotic quetiapine led her to pack on more than 40 pounds and sleep 17 hours a day. (Maya requested that her last name not be used.)

But those numbers don't do justice to her story. It's the long list of diagnoses Maya collected before she was 21, from borderline personality disorder to agoraphobia to obsessive-compulsive disorder, that begin to hint at how little we understand autism in women.

Her conversation with psychiatrist number 14 went something like this:

Do you hear things that others don't?
Yes. (Maya's hearing is excellent.)
Do you think others are talking about you behind your back?
Yes. (Maya's extended family is particularly gossipy.)

The psychiatrist didn't explain exactly what he was trying to assess. Literal to a fault, Maya didn't explain what she meant by her answers. She left his office with her eighth diagnosis: paranoid personality disorder.

Maya does have some of the conditions she's been diagnosed with over the years—she's been depressed since the age of 11, has crippling social anxiety, and in her teens, wrestled with anorexia. But these were just expressions of the autism that was there for anyone to see had they looked closer. "It's all secondary to the Asperger's," says Maya, now 24. "I get depressed and anxious because life is difficult; it's not the other way around."

It's not uncommon for young women like Maya to be repeatedly misdiagnosed. Because autism is at least three times as common in boys as in girls, scientists routinely include only boys in their research. The result is that we know shockingly little about whether and how autism might be different in girls and boys. What we do know is grim: on average, girls who have mild symptoms of autism are diagnosed two years later than boys. There's some debate about why this might be so. From the start, girls' restricted interests seem more socially acceptable—dolls or books, perhaps, rather than train schedules—and may go unnoticed. But the fact that diagnostic tests are based on observations of boys with autism almost certainly contributes to errors and delays.

As they enter their teens, girls struggle to keep up with the elaborate rules of social relationships. Cribbing style notes on what to say and how to say it, many try to blend in, but at great cost to their inner selves. Starting in adolescence, they have high rates of depression and anxiety—34 and 36 percent, respectively. A few studies have also found an intriguing overlap between autism and eating disorders such as anorexia, although the studies are too small to estimate how many women have both.

Even after a girl gets the right diagnosis, she may be offered behavioral therapy and specialized lesson plans, but they're essentially the same services offered to a boy in the same situation. Scientists and service providers rarely acknowledge the additional challenges being female may bring, whether physical, psychological, or societal. There are no guidebooks for these girls or their families about how to deal with puberty and menstruation, how to navigate the dizzying array of rules in female friendships, how to talk about romance and sexuality or even just stay safe from sexual predators. Advocates and scientists in other disciplines have run up against and resolved many of these same problems, but in autism, the fact that boys and girls are different is sometimes treated as if it's a startling new discovery.

In the past two to three years, there has been an uptick in the attention paid to the issues that affect women with autism. More money is now available for scientists to study whether and how autism differs in boys and girls. This past year, the journal *Molecular Autism* dedicated two special issues to research specifically exploring the influence of sex and gender on autism. "Almost overnight, we went from a couple of people talking about sex differences to everyone studying this as a major factor in the field," says Kevin Pelphrey, Harris Professor at the Yale Child Study Center.

Unpublished results from Pelphrey's lab confirm what common sense suggests: women with autism are fundamentally different from men with autism. Autism's core deficits may be the same for both, but when the symptoms intersect with gender, the lived experience of a woman with autism can be dramatically different from that of a man with the same condition.

Girl Power

From its first clear description in 1943 by Leo Kanner, autism has been known to crop up in more boys than girls. But why this is so remains a mystery.

At first, scientists looked for the simplest explanation: that a boy who carries a faulty stretch of DNA on his single X chromosome develops autism, whereas a girl who inherits the same mutation would be unaffected because she has a second X chromosome to compensate.

But the search for this X factor went nowhere. "I think the thinking is now moving more to the idea that women are protected, which I know sounds like two sides of the same coin, but it plays out in a different way," says Stephan J. Sanders, assistant professor of psychiatry at the University of California, San Francisco. The idea is that, for as yet unknown reasons, women can tolerate more mutations than men can, and so need a bigger genetic hit to develop autism.

A 2012 paper that laid out this "female protective effect" in autism marked a turning point in the field, bringing the topic of girls with autism into the spotlight. "Once the genetics community became interested in it, it just absolutely took off," says Pelphrey.

Around the same time, Pelphrey and his collaborators won a five-year, $13 million grant to probe the differences between girls and boys with autism, as well as their unaffected siblings. They are recruiting 250 girls with autism between 6 and 17 years old at six sites across the United States. They plan to characterize the behavior, genetics, and brain structure and function of these girls and compare these findings with data from 125 boys who have autism, as well as from 50 children in each of the following groups: typically developing boys, typically developing girls, unaffected male siblings, and unaffected female siblings of children with autism. "We're trying to address the question: Are girls different? And how are they different?" says Pelphrey.

A few studies have explored this question. There seems to be an overall consensus among scientists that at the more severe end of the spectrum—characterized by low intelligence quotient (IQ) and repetitive behaviors—there is little outward difference between girls and boys with autism. It's at the other end of the spectrum that the science is fuzzier. Given the small numbers of women with autism in the studies, there are few definitive answers.

"Clinically, my general impression is that young girls with autism are different [from boys], but it has been very hard to show that in any kind of a scientific way," says Catherine Lord, director of the Center for Autism and the Developing Brain at Weill Cornell Medical College in New York City. On average, girls are more chatty, less disruptive, and less likely to be entranced by trains or moving vehicles than boys are, she says. However, she adds, this is also true of typical girls and boys, so it becomes difficult to separate gender differences in autism from gender differences in general.

Early studies estimated that at the high-IQ end, the male-to-female ratio is as high as 10 to 1. The picture emerging from studies looking at girls with autism over the past few years suggests this ratio is artificially inflated, either because girls at this end of the spectrum hide their symptoms better, or because the male-biased diagnostic tests aren't asking the questions that might pick up on autism in these girls—or both.

"For some males, you can make the diagnosis at least provisionally in your mind within 10 minutes of them coming into your office," says Simon Baron-Cohen, director of the Autism Research Center at Cambridge University in the U.K. "Whereas for some of the women, it might take half an hour or not till halfway through a three-hour diagnostic interview before they're revealing what's behind the mask."

Hidden Hurt

It takes hours to see glimpses of the pain Maya has endured over the years. She makes eye contact, pokes fun at herself, and takes turns in conversation—things people with autism are generally known to have trouble doing. On a warm June day in London, dressed casually in a T-shirt and shorts, she looks like any other British 20-something. "You can see by meeting with me that I'm quite chatty and that people wouldn't guess that I have Asperger's," she says.

Maya is proud of her accomplishments—and rightfully so. She excelled at school: she could read fluently by age five and began reading four or five books a week. She was lead violinist at her school, performing at the Barbican Centre in London, and can also play piano and viola. She taught herself to play the clarinet, and after nine months of lessons, performed a Mozart concerto at her school.

But as the conversation turns more intimate, she and her mother reveal the agony that has formed the backdrop to her achievements. At four, Maya had severe separation anxiety and screamed every time strangers entered her nursery school. Later, at her all-girls school, she sat by herself at playtime and read everywhere, even onstage at a cousin's raucous wedding. She struggled with small talk, regularly made social faux pas—blurting out the denouement of a mystery, or reciting divorce statistics at an engagement party—and rambled on about her interests so long that her mother devised a secret gesture, a tap on the watch, to signal her to stop.

Any small disruption in her routine—dinner on the table 10 minutes later than promised, a late appointment, her little brother sitting in her favorite chair—could ruin her week. ("It's not something I like about myself," Maya says. "I can't help having this need for wanting everything to be the same—but I do.") She rarely got a good night's sleep and had debilitating nightmares. She turned down invitations to "aimless" social activities such as shopping, and called other girls out when they flouted the school's rules, turning would-be friends into enemies.

By the time she was 8, she was bullied so much at school that she became sick with anxiety every Sunday night. At 11, her parents finally switched schools, but she was bullied there as well—even on the 45-minute bus ride each way.

Looking for the common factor, Maya's logical mind pinned the blame on herself rather than on the cruel social games of girlhood. "I thought: *Everything's different—the school's different, the people are different, yet the bullying is the same,*" she recalls. *"Therefore, the only thing it can be is that something's wrong with me."*

The bullying got violent and more vicious as she got older. She recalls one set of girls telling her that the world would be a better place if she weren't in it, and that they felt really sorry for her parents. Ever honest herself, Maya believed them: "I won't say things unless they're true, so I thought, why would they?"

When she was about 12, Maya began secretly cutting herself. Like many girls with autism at this age, Maya was keenly aware of all the ways in which she was being excluded by her peers. She became intensely depressed, launching her long and dysfunctional relationship with the psychiatric establishment.

At 15, to keep herself occupied during the unstructured summer holidays, Maya began volunteering with boys who have autism—at first only because the organization was around the corner. She never made the connection that she might have something in common with them. She brought one of the young boys home to visit once, and still neither her father, a physician, nor her mother, a clinical virologist, picked up on any similarities.

"My picture of autistic was what this little boy was like—and that's not what Maya's like. He was nonverbal, disruptive," her mother, Jennifer, says. "I would not have made the connection with all the unhappiness she experiences."

The bullying stopped at 16 when Maya was moved into a new class at the school. But soon after, she became obsessed with controlling her weight. Like many other adolescent girls with autism, she developed an eating disorder. The way she sees it now, that preoccupation was an outgrowth of another aspect of her autism—her love of numbers. "I was obsessed with decreasing the number of calories I ingested, and the numbers on the scale going down," she says. Anorexia also resonated with her perfectionistic streak. "It's fine if it's something like learning musical instruments," she says. "It's not so fine if you decide to starve yourself, because I wanted to do that to perfection as well."

Over the next two years, Maya became "a master of disguise," hiding her food and exercising in secret, even on a family safari in Kenya in July 2009. "You know what I remember about that trip? I remember that I gained four hundred grams in two weeks; that's what I remember," Maya says.

Each accomplished target led to the next until at one point Maya, who is five feet six and a half inches, weighed just under 44 kilos (about 97 pounds). "The anorexia has been, from my perspective, possibly the most difficult thing to cope with, out of all the things we have gone through," says her mother, Jennifer.

In August 2009, relenting to her parents' pleas, Maya went back to her first psychiatrist. She emerged with six diagnoses, including anorexia, generalized anxiety, bipolar disorder, and agoraphobia.

In October of that year, despite the ongoing anorexia, Maya's parents drove her to Cambridge University, her life's dream until that point, crying all the way home because they were so worried about her. At first, Maya seemed to thrive—she enjoyed her classes, and made friends who were "quirky" like her, her mother says. But soon, she stopped talking about her new friends, and when her friends would knock on her door, she simply wouldn't answer. The depression that had come and gone since she was 11 resurfaced. "I didn't want to socialize, I didn't want to see anyone, it was too difficult," Maya says. She also began taking overdoses of her meds, enough to get her on the radar of the local mental health team.

Maya's second year was the same. She continued to struggle with anorexia: "It clocked that my goal was to weigh nothing." Then one day, her counselor at Cambridge pointed out that even if she had no fat or muscle, she would still carry the weight of her bones. "Therefore, I could never weigh nothing, even if I was dead," Maya recalls thinking. As is the case for many people with autism, facts hold great power for Maya. The logic of the counselor's statement got through to her like no amount of pleading from her parents had. "I realized that what I was doing was completely pointless. I was never going to get where I wanted to."

The relief from the decision to stop controlling her weight carried Maya through her second year. The family once again went on an exotic holiday, this time to the Galápagos, and Maya seemed at peace. She swam with the dolphins—and she ate.

But back at Cambridge for her final year, she again sank into a deep depression. Her mother, who had rented an apartment in town and slept on Maya's floor one or two nights a week, urged her to leave university so she could focus on feeling better. Quitting went completely against the grain for Maya. "I don't give up on things," she says. "I hate it when plans change. My plan was to finish school, go to university, graduate. My plan was not to get so depressed that I had to leave university." But four weeks into the term, after getting no help from a university psychiatrist (the one who allotted her seven minutes), she made the difficult decision to leave.

Far from making her feel better, however, leaving Cambridge made her feel as if she had no future. Overweight and sluggish, she slept through her days at her parents' house. On the 29th of December, after going out to lunch (which Maya finds stressful), cooking her family dinner (which she loves to do), and a pleasant and unremarkable night of watching television with them, Maya took more than 30 tablets of paracetamol (acetaminophen), about 15 codcinc pills, and all the quetiapine she could lay her hands on. "Nothing was getting better," she says. "I just gave up; I'd had enough of life."

A short while after taking the pills, Maya panicked that she was still awake, and that she might begin to vomit, something she dreads. She woke her parents and, within a half hour of arriving at the emergency room, fell into a coma.

Social Networks

Social isolation, bullying, and depression are not exclusive to girls with autism—boys experience them too. But for older girls with autism, the intricacies of their social world add layers of complexity.

In early childhood, boys and girls with autism are about the same. If anything, girls appear to be more social—whether because they actually are or are just perceived to be. As they edge closer to adolescence, however, girls with autism lose this early social advantage, becoming less and less likely to have friends, and more likely to be isolated. "It can be very, very tough for them," says Pelphrey.

For some girls, that may be a result of having mostly been in classes with boys who have autism. But even for girls who are placed in mainstream schools, the rituals of female adolescence can be boring or bewildering.

Adolescent boys tend to socialize in loosely organized groups focused on sports or video games, allowing a boy with minimal social skills to slide by, says Kathy Koenig, associate research scientist at the Yale Child Study Center. "For girls, socialization is all about communication, all about social-emotional relationships—discussions about friendship, who likes who and who doesn't like who and who is feuding with who," Koenig says. "Girls on the spectrum don't get it."

Adolescence can be a confusing time for any young girl, but for a girl with autism, "trying to make friends and not understanding why the friendships aren't lasting, or why you're not being included when people are making plans" can be incredibly isolating, says Baron-Cohen. "You're aware enough to know that you're failing, basically."

Ostracized and aware of it, adolescent girls with autism become highly anxious and depressed, and many develop eating disorders. This trend remains constant until late middle age, when clinicians suspect that, as they are known to do in the general population, the differences in mood disorders between men and women with autism may even out.

There are any number of programs for people with autism that teach specific behavioral skills—improving eye contact or turning your body toward the person you're speaking to, for example. But there is almost nothing to give adolescent girls the kind of emotional support that only comes from true companionship.

In the United States, there seem to be just three such programs—one at Yale, one at the University of Kansas, and a new center in New York City.

The Yale program, which Koenig launched more than three years ago, brings girls with autism together for yoga, or to make jewelry or to watch the blockbuster movie *Frozen*—the same kinds of activities typical girls might do. There are different groups for young girls, teenagers, and young women, with about 102 families registered in total. Some groups are purely social, but others offer training for interviews, or provide support for women in college.

The Kansas program, called Girls' Night Out, goes one step further by pairing typical girls and girls with autism. Groups of girls might visit a hair or nail salon, a coffee shop or gym, or learn how to buy clothes appropriate for their age and the weather.

"I was worried at the beginning that people would think I was trying to change them, that I was focusing on appearance," says program director Rene Jamison, clinical associate professor at the University of Kansas in Kansas City. But hearing from parents and from the girls themselves what a difference it has made to their confidence levels has been reinforcement enough, Jamison says.

Learning to brush their own hair or teeth and to use deodorant can make all the difference to teenage girls in social situations, Jamison says. "These are skills that other girls are picking up on naturally, and getting better at," she says. "That's not happening naturally for some of the girls we work with, and so, just like social skills, it has to be an explicitly taught thing sometimes."

Growing Up

Even with early diagnosis, with social skill and behavioral training and numerous other avenues of help, girls with autism and their families have little help coping with a key milestone: puberty.

Isabel Haldane, or Lula, as everyone calls her, is 11, and for most of her life has had multiple experts dedicated to helping her navigate the world, beginning with her anthropologist parents. Until she was about 15 months old, Lula seemed precocious, walking early and rapidly picking up words. Sometime between 15 and 18 months, she lost her words and began humming—the closest approximation of the sound, her mother says, is in the movie *Finding Nemo* when the character Dory is trying to imitate a humpback whale—during the day, and wailing in frustration all night. She also didn't make eye contact or respond to her name, so by age 2, she was diagnosed with autism and recruited into an early-intervention program.

Since then, Lula has had combinations of speech therapy, playtime therapy, pivotal response therapy—a form of applied behavioral analysis, the most common autism treatment—occupational therapy, and social-skills training. Starting at age 3, she placed into her local public school in suburban Connecticut, where she spends 11 months of the year, but she still has therapists who work with her for about five hours a week at school, and another hour a week at home.

Thanks to all this help, by age 5 Lula was mostly toilet trained and began to talk. By 9, she began sleeping through the night, and her parents could finally stop taking turns staying up with her all night. She scores below average on traditional IQ tests, but like many children with autism, she is adept at some things and stumped by others. She can shower, dress herself, pack her bag, and wait for the school bus at the bottom of her parents' driveway, but she might do it all at 5:00 a.m., hours before she's supposed to. She can decode any word—"catastrophe," for example, or "encyclopedia"—but ask her what the word means and she might respond with "I love Scott Walker" (a classmate, not his real name).

Many of the school's students have known Lula since she was 5. But while the other girls have moved on to dance and gymnastics and music recitals, Lula is still mostly fixated on Hello Kitty. As kind as the girls are to Lula, they see themselves more as her protectors than as her friend.

Lula feels any social rejection acutely. She has memorized the birthdays of all of her friends but knows she is only invited to two parties a year. On a recent afternoon, as she arranged and rearranged her Hello Kitty–themed room, she perseverated about not seeing her friends at camp and about not wanting to get older.

Lula's periods began just before she was 10, and she is fully developed physically, a beautiful brown-haired girl who looks older than her years. Puberty has brought enormous unforeseen challenges. Although Lula has learned how to use sanitary pads and sometimes remembers to change them, she doesn't always think to dispose of them properly. "I didn't even realize how much instruction it took to deal with a monthly occurrence. I didn't know where to go on the Internet or who to ask," says Hillary Haldane, Lula's mother. "Where is the tutorial on this?"

Lula also shows a preteen's healthy curiosity in sex, but none of the embarrassment or hesitation that might typically accompany it. A boy at school Lula has taken a shine to comes up in conversation often. She might announce that she wants to touch his penis, or smell his crotch. When she has blurted out these comments in school, her teachers' reaction has been to isolate her. Knowing it goads the adults around her, Lula has taken to doing it even more.

"As we go into middle school, this is the biggest fear I have: her saying these things and then being ridiculed or bullied for it," says Haldane. Even more worrying for her parents is the sort of attention she might attract outside school: "It's so terrifying with the sexual predatory behavior that she might face, especially because her body is quite developed, and her sexual curiosity, and how much more I have to consider what her behavior signals to others as opposed to if she was a boy on the spectrum."

Deeper Worries

Safety is an enormous concern for women who cannot advocate for themselves, and it weighs heavily on families' minds. For Karleen, whose daughter Leigh, 28, is a nonverbal woman with autism, fighting for her daughter's dignity has become nearly a full-time occupation. (Karleen asked that her and Leigh's last names not be included, to protect Leigh's privacy.)

Leigh uses a few words but, for the most part, cannot follow commands or speak. The youngest of three siblings, Leigh, like Lula, lost speech at 15 months and was diagnosed at 2. But she is unable to care for herself at all, and because of her tendency to hurt herself and others, needs around-the-clock care. "When she's anxious, Leigh can strip right down. She can be trapped that way buck-naked, until she can get the anxiety under control," says Karleen. When she has her period, Leigh's anxiety can skyrocket so much that she might shred pads into tiny pieces.

After years of searching, Leigh's family, based in Belmont, Ontario, found her a residential program that created the kind of calm and routine that Leigh needs. But the agency must follow union guidelines on equal employment, meaning that it might pair Leigh up with a male attendant.

For the past two years, Karleen has been appealing to officials at every level of the agency to allow only female attendants to work with her daughter—to no avail. In fact, she says, the agency may have to refer Leigh elsewhere because it cannot afford the legal fees to explore whether the law would allow it to only hire female attendants.

A former public health nurse who worked in women's shelters, Karleen is only too aware of the potential for abuse, particularly with male attendants. "I think this could be a huge issue in the future," Karleen says. The equal-employment-opportunity law was meant to protect people's rights, Karleen says, but is paradoxically harming women like Leigh who need support and cannot advocate for themselves. "If you are able-bodied and you can speak or you can gather support, then you can challenge that or work that legislation on your own behalf, but if you're someone like Leigh, then how can you be protected?"

Different Worlds

Whether it is Leigh's thorny legal situation, Lula adjusting to her budding sexuality, or Maya's run-ins with psychiatrists who misunderstood her pain, the issues that dog women with autism have everything to do with their gender. For the first time, scientists are beginning to incorporate what they know about typical girls and their social world to understand girls with autism.

For example, it's been known for decades that boys' and girls' social worlds are starkly divergent and that they learn the rules to function in these worlds in disparate ways. "There's really good data to show that in typical girls and boys, the socialization trajectory is different," says Koenig. "People never took that into account when they're studying autism."

The multisite project that Pelphrey leads is making headway into learning how girls with autism are different—both by recording their behavior and by scanning their brains. For example, one of the cardinal observations about autism is that people with the condition seem uninterested in, or at least disengaged from, social interactions. Intriguing brain-imaging evidence from Pelphrey's lab suggests that this is true only for boys with autism.

"The most surprising thing—it might not be surprising to the clinicians out there, but to the scientists—is that we're seeing strong social brain activation or function in girls with autism, which is, strictly speaking, counter to everything we've reported ourselves and other groups have reported," says Pelphrey. "Their social brains seem to be intact."

The social brain is an interconnected set of brain regions, including the face-processing fusiform gyrus; the amygdala, an emotion hub; and the superior temporal sulcus, which tracks other people's attention and movements. Imaging studies have reported that the social brain is underactive in people with autism, but Pelphrey's lab has found that if typical girls have the most active social brains and boys with autism the least active, typical boys would tie with girls who have autism somewhere in the middle. "That kind of blew us away," he says.

Particularly interesting is the unpublished observation that in girls with autism, the social brain seems to communicate with the prefrontal cortex, a brain region that normally engages in reason and planning, and is known to burn through energy. It may be that women with autism keep their social brain engaged, but mediate it through the prefrontal cortex—in a sense, intellectualizing social interactions that would be intuitive for other women.

"That suggests compensation," Pelphrey says. It also jibes with women like Maya saying they have learned the rules of social interactions, but find it draining to act on them all day. "It's exhausting because it's like you're doing math all day," Pelphrey says.

Pelphrey is right that this finding isn't entirely a surprise to clinicians. Some scientists who regularly see women with autism have picked up on their remarkable ability to learn the rules enough to camouflage their symptoms—the way Maya has learned to. ("I don't like making eye contact," Maya says. "I do it because I have to and I know it's appropriate.")

This means clinicians have to be more creative when diagnosing women on the spectrum, rather than simply looking for, say, repetitive behavior, as they might with men. "Without their self-report telling you how stressful it is to maintain appearances, you wouldn't really know," says Francesca Happé, director of the MRC Centre at King's College London. "They have good imitation, good intonation in their language, body language—surface behavior isn't very useful for a diagnosis, at least for a certain set of women on the spectrum."

Overall, the concept of compensation in women with autism hasn't been well studied, Happé says. Compensation could be cognitive—learning the rules intellectually rather than instinctively, as Pelphrey describes it—or social, such as learning to mimic others. There are also societal factors at play. "Are we more tolerant, at least in some Western societies, of a girl who is very, very quiet and socially aloof, compared to a boy? I don't know; I suppose you could say we have higher expectations of women," says Happé. "All of these are hypotheses and they're only interesting if they're testable."

A few teams, including those led by Happé and Baron-Cohen, are trying to find ways to get behind the masks. Baron-Cohen's group is developing what he calls a "faux pas test." If a woman is getting by learning social rules one rule at a time, as Maya has, she's bound to make a lot of mistakes, he says, because she's likely to encounter a situation for which she hasn't yet learned the rules. Happé is similarly creating tests based on real-life scenarios in which her team asks women not only why somebody said something, but also what they themselves would say next. "That really trips people up. It would require them to, on the spot, get it," she says.

Baron-Cohen, Happé, and others caution, however, that in some cases, women may have learned to cope enough that they don't actually need a diagnosis.

"If they're coping, do they want to think of themselves or for others to think about them in that way?" asks Happé. "Then it becomes a big ethical issue, doesn't it?"

New Understanding

In Maya's case, learning she is on the spectrum took some getting used to. But she says she's very glad to have an explanation now for all of the difficulties she thought were unrelated to one another.

After she came out of her coma, Maya spent a week in intensive care and nine weeks in a terrifying psychiatric unit with severely ill patients. One threw a boiling cup of tea at a nurse, and another head-butted a nurse so hard that her teeth went through her lip. In the early days, Maya deliberately burned her arm with the hot water available for making tea, and threatened to try to kill herself again as soon as she got home.

But as the weeks passed, she started to feel better. She was given an antidepressant that seemed to work for her, and she lost the weight she had gained when taking quetiapine. She met a young woman who has since become her best friend. Then, several months after she left the hospital, she got the autism diagnosis.

After her disastrous encounter with the psychiatrist who decided she had paranoid personality disorder, a doctor who had been kind to her while in the hospital offered to take Maya back as a patient. It was only when Maya began complaining about the ridiculousness of offices being closed on "bank holiday" Mondays ("Weekdays are for work!"), and how overwhelming it was for her to walk down a noisy street, that the psychiatrist added up the signs to arrive at the correct diagnosis.

A full 18 months after Maya came home from the hospital, she went back to Cambridge for her final year and switched her focus from genetics to psychology and cognitive neuroscience. She burst into Baron-Cohen's office at Cambridge one day while he was in a meeting, announced that she has Asperger's syndrome, and asked if he would supervise her dissertation on mirror neurons and autism. He agreed. She still has bouts of depression, but her stay in the hospital taught her how and when to ask for help. "When I came out of hospital, I basically lived along the lines of 'If it's stressful, don't bother doing it,'" she says. "Nothing is worth getting that depressed."

The university accommodated her diagnosis, allowing her to take her exams alone and with breaks in between, and in June 2014, despite some ongoing depression, Maya graduated from Cambridge. "If you can go in two and a half years from being locked in a psych unit to graduating from Cambridge, you can do anything, really," Maya says.

After graduation, Maya worked for a year at a local primary school, supporting boys with autism in the class. She didn't tell the school she has autism, and successfully held down the job all year. She enjoyed it so much, in fact, that last month she began training to be a primary school teacher, specializing in mathematics, and plans to either teach mathematics or work with special-needs children. And this time, Maya revealed on the application form that she has autism. "She agonized about it a lot; she didn't want people to prejudge her," says Jennifer.

Outside of her teacher training, Maya spends time with her best friend, even going on a holiday together "with massive success," and has dated men on the spectrum. Most of all, she is committed to learning how to take care of herself the way only she can. One day this summer, she went on a "fun run"—"which as far as I'm concerned are two words that should not be put in the same sentence," she says—a loud and colorful obstacle course that Maya researched thoroughly online and prepared for with earplugs. When she has a bad day, she has learned to wind down with multiple episodes of *Grey's Anatomy,* which she has watched enough times to be able to fake being a doctor. She and her brother now laugh about her need to sit in the same seat at the dinner table, and her parents have learned to respect her need for solitude, despite their fears about what she might do when alone.

She has also been talking about her autism—at town council meetings, to groups of teachers and trainee therapists—and helping to train the staff at doctors' offices to accommodate people with autism's need for order and quiet.

Maya still gets depressed, still rarely has a night free of nightmares, and may still go into a tailspin if her routines are disrupted. But she is better than she was at asking for support—and often gets it from a therapist who specializes in autism whom she sees every other week, or more often if necessary.

"The more I understand myself, the more I can explain to other people what I find difficult, and the more they can help me," she says. "Life isn't easy for me, but I understand myself so much better now."

CHARLES C. MANN

Solar, Eclipsed

FROM *Wired*

A FEW MINUTES AFTER I meet E. V. R. Raju, a vision pops into my head. I can see him on one of those lists of the World's Most Important People released by the likes of CNN, *Forbes,* and *Time.* Besides the obvious entrants like the president and the pope, the lists always also include a few buzzy, click-generating names: Emma Watson, perhaps, or Bono. Raju is certainly not in either of those categories. He is the environmental manager of a coalfield in northeastern India.

The Jharia coalfield, where Raju works, is India's biggest and most significant, covering some 170 square miles. It has been on fire, calamitously, since 1916; entire villages have collapsed into the smoking ground. Raju's job is to put out the fire, so that his company can roughly double the mine's output in the next five years. Whether—and how—he can perform this task will have much more effect on the future of the world than anything, with all due respect, likely to be accomplished by UN-addressing actresses or aging Irish rock stars. In other words, if one were compiling a list of the World's Most Important People, Raju should be on it.

To judge by my visit, Raju is a busy guy. A line of functionaries with documents in envelopes wait outside the door of his surprisingly small office. Saying he has little time to talk, he waves aside a minion who offers to bring in tea. "The prime minister said the fires have to go out," he tells me. "He said money was no issue. He made a statement a few days back. Things have to happen fast." As I scribble notes, it occurs to me that a list of Most Important

People should also include Prime Minister Narendra Modi—who just might deserve the top spot.

For two decades, Americans have been barraged with news about the ascent of Beijing—its economic power, its enormous size, its rising voice in world affairs. Much less attention has been paid to New Delhi. This will change. Already earth's fastest-growing major economy and its biggest weapons importer, India is on track to become the world's most populous nation (probably by 2022), to have its biggest economy (possibly by 2048), and potentially to build its biggest military force (perhaps by 2040). What China was in the American imagination in the 1990s and 2000s, India will be in the next two decades—a cavalcade of superlatives, a focus of fears.

Nowhere is this truer than on climate change, tomorrow's single greatest challenge. For years, attention has focused on the role of China, the largest emitter of greenhouse gases, and the United States, one of the largest per capita emitters. In November 2014 the two nations promised substantial limits on greenhouse gas emissions for the first time; China has pledged that its carbon dioxide output would fall after 2030, while the United States has vowed to cut its output by more than a quarter in about the same time frame. Indeed, China's emissions have fallen so fast in the past year that many believe it may achieve its target ahead of time—the biggest stride yet in the fight against climate change.

India's carbon output, by contrast, is growing faster than any other country's. Should that trend continue—and there is reason to think that it will—India could surpass China in 25 years to become the world's greatest emitter. Conceivably, its increasing emissions could offset all the efforts at curtailment in the rest of the world, leading to catastrophe. "India is the biggest piece of the puzzle," says John Coequyt, Sierra Club's director of federal and international climate campaigns. "Is there a way for that rapid growth to happen quickly and pull people out of poverty using a lot more renewable energy than has ever been used before? Or will they build more of what they have—huge coal plants with almost no pollution controls?" The latter course, he says, would be "a disaster for everyone."

The inevitable conflict between India and other nations could come to a head as early as December's international climate talks in Paris. India appears to be participating only reluctantly—it was the last major nation to release an emissions plan. Although the plan projected big increases in solar and wind power, energy efficiency, and reforestation, it didn't actually promise to cap greenhouse gases. It also demanded rich nations pay for most of the cost, which it estimated to be "at least $2.5 trillion . . . between now and 2030"—more than $166 billion a year for the next 15 years. Within weeks, environmental groups were complaining that India was threatening to capsize the negotiations, holding the whole world hostage to its demands.

Matters look different inside India. There, officials and academics have long argued that Western nations are demanding that India industrialize without burning even a fraction of the fossil fuels that developed nations consumed when they industrialized. And Indians resent that Western nations insist on the right to judge Indian performance while refusing to help with the cost of transition. "The West—not India—filled up the air with carbon dioxide," says Sunita Narain, director general of the Centre for Science and Environment in New Delhi. "The West is asking *us* to pay for *its* mistakes. They are saying, 'Oh, you are a rich country now, you can cover the cost.'"

A "premature superpower," in the words of economics writer Martin Wolf, India is focused on both increasing its influence abroad and raising its living standards at home. Its per capita income is just $1,778. (The comparable figure for the United States is $51,013; China's is $6,050.) Even India's wealthy are poorer than their counterparts in the West; of the nation's richest 10 percent, a third live in households with no refrigerators. Worse, some 300 million Indians—a quarter of the population—have no electricity at all. Nearly as many have only intermittent access to it. Most of these people use kerosene for lighting and cook their food on wood or dung fires. The smoke kills about 1.3 million Indians a year, according to the World Health Organization.

Providing power to these literally powerless people is "a priority in every imaginable way—human, economic, and political," says Navroz Dubash, a senior fellow at the Centre for Policy Research in New Delhi, who is a lead author of reports for the UN's Intergovernmental Panel on Climate Change. Partly in consequence, India's demand for electricity is widely expected to double by 2030. The Modi government is determined to satisfy that demand. In fact, Modi—arguably the most powerful Indian prime minister in three decades—is pursuing this goal by charging down not one but two paths, each fraught with difficulties.

The course touted most by outsiders is an aggressive program to expand solar power. In his former position as chief minister of the western state Gujarat, Modi oversaw the construction of Asia's biggest solar park, a giant utility with battalions of solar panels. Soon after being elected prime minister in 2014, he announced that India would produce 100 gigawatts of solar power by 2022 (the United States now has about 20 gigawatts). Earlier this year, India unveiled plans to build the world's biggest solar park, in the northern state of Madhya Pradesh. This path is next to impossible: no nation has ever expanded its renewable-energy infrastructure at the speed Modi envisions. India could easily spend huge sums and still fall short of its ambitions, leaving tens of millions of people in the dark.

Simultaneously, Modi is forging a second, contradictory path: to power the nation using India's vast coal reserves, among the top five in the world. Increasing output will require transforming the corrupt, hidebound state enterprise Coal India and moving as many as a million people out of the way to extract the coal. To generate electricity from it, India plans to build 455 new coal-fired electric power plants, more than any other nation—indeed, more than the United States now has. (India's existing 148 plants, which provide two-thirds of its electricity, are among the world's dirtiest and most inefficient.) This strategy has a brutal downside: vastly increased carbon emissions that would make it nearly impossible to prevent global temperatures from rising more than 3.6 degrees Fahrenheit, the goal of the Paris talks. Higher temperatures will have catastrophic implications around the globe—and India, with its long coastline, scarce water supplies, and hot climate, may be more vulnerable to the effects of climate change than any other big nation.

Last summer I spent three weeks in India, speaking to academics, activists, businesspeople, and politicians concerned with the nation's energy and climate issues. Not one person believed that India had the financial muscle to pursue both paths. One will have to be downsized, or even abandoned. In practical terms, the nation will end up making a choice: more coal or more renewables. That choice will affect the lives of the hundreds of millions of Indians who today live without lights, refrigerators, air conditioners, telephones, or the other necessities of modern life. But its ramifications will also ripple across the rest of the world.

"Indians used to be furious at the way decisions in the West—decisions in Washington and London they had no say in—could upend their lives," Narain says. "Now, I sometimes think, people in the West will understand what that feels like."

From an airplane window, the coastal state of Gujarat seems like a monument to the ambitions of its native son, Narendra Modi. In a former badlands 100 miles from Ahmedabad, its biggest city, I could see sunlight reflecting from the Charanka solar park, Asia's biggest. Dozens of rectangular photovoltaic arrays, regular as midwestern wheat fields, were scattered in a broad U over a mile on each side. By squinting a little I could talk myself into thinking I saw power lines spiderwebbing from the arrays, bringing hundreds of megawatts out of an otherwise barren expanse. Twenty miles from the airport was a metallic ribbon, a half mile long and over a hundred feet wide: a solar park built atop an irrigation canal. Southeast of the city was a second, a 2-mile tunnel of aluminum and polymer. As the plane approached the tarmac, solar panels stood like sentinels atop buildings everywhere—a vision of a green future, almost all of it brought into being by the preternaturally determined Modi.

The center of India's oldest civilizations, Gujarat is at once a cradle of Hindu identity and a busily cosmopolitan place, full of traders from across Asia. Modi arguably represents both traditions, the insular and the global. Like a subcontinental Bill Clinton, he is a charismatic figure with a resonant origin story, a passion for politics, and a reputation for flexible ethics. Modi was born in 1950, the son of an impoverished tea-stall owner in a remote Gujarat town. From adolescence, he worked as an operative for the Rashtriya Swayamsevak Sangh (National Volunteer Organization), a nativist outfit dedicated to the idea that India is an essentially Hindu nation, founded on Hindu beliefs and ideals. It has a network of schools, charities, and clubs run by disciplined cadres of conservatively dressed activists—and a violent aura; it has repeatedly been accused of organizing attacks on Christians, Muslims, Sikhs, and other non-Hindus.

In 1987 Modi joined the Bharatiya Janata (Indian People's) Party, a pro-Hindu, nationalist party tied to the RSS. He rose steadily and won election as chief minister in October 2001. A few months after the vote, a Gujarat train loaded with Hindu pilgrims and activists caught fire, killing dozens of passengers. Angered by rumors that the blaze had been set by Muslims, club-wielding Hindu thugs murdered a thousand or more people, most of them Muslim. Human rights groups charged that the BJP had encouraged the attacks. Modi, they said, stood by as Muslims died. An inquiry dismissed the accusation, but the riots stained his reputation; in 2005 he became the only person ever denied a U.S. visa for "severe violations of religious freedom." (The decision was reversed in 2014.)

Alarmed by the fallout, Modi shifted gears, refashioning himself as a sharply dressed, tech-friendly progressive who lured major companies, foreign and Indian alike, to invest in Gujarat. He also became one of the world's most prominent advocates for solar power. In a "green autobiography" published in 2011, Modi promised to transform hot, dry Gujarat, with its 55 million people, into a model of sustainable development, simultaneously increasing irrigation and recharging aquifers, converting hundreds of thousands of cars and trucks from gasoline to natural gas, and turning the state capital, Gandhinagar, into a "solar city." He created Asia's first ministry of climate change and led a pioneering program to install solar panels atop irrigation canals, shielding the canals from evaporation and generating power without covering scarce farmland. "I saw more than glittering panels," said UN secretary-general Ban Ki-moon, inaugurating a canal-top project in January. "I saw the future of India and the future of our world."

Getting to that tomorrow will be difficult. During my visit to Charanka, it was about 110 degrees and windy. Dust, whipped into the air, obscured the sun and coated the solar panels. Pipes beneath the arrays carried water to wash them. Solar parks, farms for electrons, effectively had to be irrigated. Here and there the serried lines of panels wobbled, nudged out of alignment by harsh conditions and land subsidence. Energy from the sun today is responsible for about 1 percent of India's electricity; even in Gujarat, it amounts to just 5 percent. Optimistic government scenarios show its share rising to 10 percent by 2022. The state-owned Power Grid Corporation of India has proposed creating huge installations in Indian deserts to increase solar's share to 35 percent by 2050. Little I saw in Charanka reassured me about the plausibility of these goals. Not one person I contacted at the park would speak to me on the record; Gujarat Power, the state-run developer of the project, had stopped issuing triumphant press releases. (Gujarat has quietly junked its climate action plan.) Perhaps the lack of interest in accommodating foreign journalists meant nothing. But the complete silence when I asked about the other part of solar power—energy storage—seemed to speak volumes.

Solar panels generate electricity only between sunrise and sunset—from about 6:45 a.m. to 6:45 p.m. during my visit. To provide electricity at night, power generated in daylight must be stored for later use. Typically storage systems employ the sun to heat a liquid (water, say, or molten salt); at night the stored hot liquid drives a steam turbine, producing electricity. In 2010 India announced seven solar-energy storage projects, one of them in Gujarat. Only one, in another state, has been built. The others were abandoned when the builders discovered that the air is so hazy, their initial estimates of potential solar power were off by as much as a quarter.

Renewable-energy advocates are surely correct that these difficulties can be solved with sufficient will and money. That's why so many of them cheered Modi's election as prime minister in May 2014. Extolling Hinduism's ancient environmental beliefs, the BJP promised in its election manifesto to "put sustainability at the center of our thoughts and actions."

A month after his election, Modi pledged he would deliver electricity to all Indians by 2019. Soon after, he moved the date to 2022. But to accomplish that, Modi about-faced, increasingly emphasizing coal. That September he conspicuously skipped a UN climate summit. The same month, the man whose autobiography denounced the "carefully orchestrated campaigns" to foment skepticism on "whether or not [climate change] was actually happening" told an audience of schoolchildren, "Climate has not changed. We have changed . . . God has built the system in such a way that it can balance on its own." In November that year, he announced India would double coal production by 2019. By then, he said, India would be producing a billion tons a year.

Eighteen hundred feet below the surface, the ancient coal-mine elevator opens into a space lined with icons of Kali Ma, goddess of the hungry earth, the deity most important to miners. Railway tracks march into the distance, disappearing in the haze. I am standing in the Moonidih mine, one of 23 mines in the Jharia coalfield in northeast India. The air is hot and intensely humid despite heroic efforts at ventilation. Forty minutes' walk away is the mine face, black and glittery in workers' headlamps. A giant bore with a six-foot drill crunches into the wall with shocking ease. Streams of water play on the head to prevent the coal dust from igniting. Wet black shrapnel flies everywhere. Behind the machine is a series of conveyor belts, rumbling one after another, conducting a black stream of coal rubble to a bunker almost four miles away.

The massive coalfield is owned by Bharat Coking Coal Ltd., a subsidiary of Coal India, one of the nation's biggest companies. Coal India owns more coal reserves than any other corporate entity in the country. Still, Jharia and BCCL occupy a special place in India's future. In addition to being Coal India's biggest colliery, Jharia is the nation's most important domestic source of prime coking coal, the hard coal that is an integral part of steel production—it provides both the necessary heat and the carbon that makes steel strong. Because any imaginable path of development involves making massive amounts of steel, ramping up production at Jharia is a top national priority. Achieving Modi's billion-ton target, company officials tell me, will require the colliery to increase its output by about 15 percent a year.

The men and women who must accomplish this huge task work in a landscaped headquarters that during my visits is full of people standing around in hallways and lobbies without obvious purpose. One morning I interview an able young engineer. Jammed into the other half of his office are a half-dozen older men, one of them his supervisor, drinking tea and telling stories. The interview lasts nearly two hours. During that time the other men do not move. Phones do not ring. Email alerts do not ping. Keyboards lie untouched. The office door opens only to admit flunkies with tea on a tray. Leaving the engineer's office, I wonder if the activists who protest India's coal-expansion plans would be comforted by this scene. Increasing productivity is going to be no easy task.

The difficulties are not all internal. The Jharia coalfield has been on fire for a century, consuming and ruining huge amounts of coal and continuing to imperil dozens of villages. When I visit the area one evening, toxic fumes, issuing from cracks in the earth, wreathe the buildings and the black, leafless trees. Patches of smoldering red are scattered like watching eyes across the charred landscape: Mordor without the Orcs.

When the coalfield opened in the late 1800s, people who wanted work simply moved into the area around the mine. In legal terms, they were squatters, but nobody wanted to drive away the work force. In time the city of Dhanbad—population about 2.7 million—grew atop the eastern end of the deposit. Dhanbad is no squatters' camp; it is a bustling, relatively prosperous city, complete with grocery stores, restaurants, middle-class apartment blocks, and bird-stained statues of dead Indian notables. To increase production from Jharia, BCCL will not only have to put out the fire, buy millions of dollars' worth of new drill-and-conveyor assemblies, and stabilize the land riddled by fire, it will also have to relocate a large fraction of this city, its satellite communities, and the burned villages in the next few years.

Because India is a democracy, people can resist such government plans. The de facto leader of the local anticoal movement is a middle-class businessman named Ashok Agarwal. A member of the Dhanbad chamber of commerce, Agarwal lives in a pleasant two-story structure built by his grandfather. His machine-parts business is on the ground floor; his struggle against BCCL, which has lasted through 20 years of protests and litigation, is headquartered in his home, amid patterned rugs, cheerful paintings, and photographs of family members. Indian law requires that BCCL relocate not only the villagers already displaced by fire but all the people who will be affected by the mine's expansion, he tells me. "It's seven hundred thousand families," he says. "More than two million people." I ask if the Indian government has ever constructed an entirely new city of that size overnight. "I don't think *any* government has," he says. "When they talk about doubling coal output, they don't mention this part." The part about moving an entire city? "Yes—that part."

Similar efforts must occur in many other places in India to fulfill Modi's goal. Unfortunately, about 90 percent of Indian coal is not Jharia-style coking coal but low-quality, highly polluting thermal coal. Outdoor air pollution, most of it due to coal, is already responsible for 645,000 premature deaths a year, according to a study published in *Nature;* New Delhi, ringed by coal plants, is said to have the world's most polluted air. Burning more coal will only make the situation worse. Already India has a high rate of chronic respiratory disease. "Success would be a disaster," Agarwal says to me. "I don't see how they get to a billion tons."

Even the smallest Indian villages I've seen have a store or two, and Luckman, in the southern Indian state of Karnataka, is no exception. At the edge of town stands a single kiosk, no bigger than an old-fashioned U.S. city newsstand. Basic supplies fill its unpainted shelves: rice, lentils, oil, chickpeas, bidis (hand-rolled cigarettes made by wrapping leaves around tobacco flakes). At night it has Luckman's only electric light: a 6-watt LED lamp, powered by what looks like an old car battery. From the battery dangles a cable that leads to the kiosk roof, on which sits a battered solar panel about the size of a cafeteria tray. This is what solar power looks like in much of rural India.

When I walk over at about 8:00 p.m., the owner is asleep with his head on the counter. Still, the store is open—the illumination allows him to keep the kiosk going after dark. Behind the clerk, a small girl crouches on the floor, doing homework in the pool of light. And behind her is an old woman, methodically rolling bidis for sale. The extended hours, the ability to do homework after chores, the chance to earn extra income—all of it comes from a single light.

Enabling even this small amount of electricity has long been a struggle. India's villages can be astonishingly remote by Western standards; a hamlet may be only 50 miles from a city but next to impossible to reach, especially when the rainy season makes roads impassable. Stringing and maintaining transmission wires in such circumstances is a nightmare.

In network jargon, India has a last-mile problem, referring to the way that bottlenecks are often found in the link that physically reaches the customer's premises. Because of this challenge, the cost of building India's electrical grid was so high that rural farmers often couldn't afford to pay for their connection. To solve the problem—and to shore up sagging popularity among poor voters—the government launched a program in the late 1980s to provide free power to low-income tribal families. Unfortunately, over time and at great expense to utilities, the benefits of the program were mostly captured by wealthier, more politically powerful families. Today, 87 percent of Indian household electricity is subsidized, but less than a fifth of the subsidies go to the rural poor for whom they were intended, and the utilities have little incentive to spend what it would take to connect them. Even if India floods the sky with coal smoke, the 300 million Indians without power still might not get connected—the worst of all possible worlds.

Enter Harish Hande. Born in 1967 and raised in the eastern Indian state of Orissa, he won a scholarship and obtained an engineering PhD at the University of Massachusetts, Lowell. His dissertation focused on rural electrification. When Hande returned to India, he went to the southern city of Bangalore, where he bought a solar home-lighting system with the last $300 from his scholarship. He sold it, installing the system himself. The transaction earned Hande enough to purchase a second system, which he sold, and then a third. He found a U.S. partner who helped him obtain additional funding. In 1995 the two men incorporated a for-profit business, the Solar Electric Light Company—SELCO. As Hande slowly built up his customer base, he kept asking villagers why they didn't already have electricity. For decades they had been waiting futilely for government agencies to fulfill promises to provide power. Why couldn't they go out and just get it themselves by installing solar panels?

According to SELCO technical manager Jonathan Bassett, the single biggest problem was financial: classically risk-averse loan officers at local banks found ways to avoid lending money for solar projects. Hande and his team came to believe that the route to India's energy future ran through the offices of low-level bank functionaries. Persuading and cajoling, experimenting and testing, they gradually installed 300,000 solar-power systems in remote villages in southern India and Gujarat, along with 45 branch offices to provide service and maintenance. As a rule of thumb, Bassett tells me, "We won't install systems without a branch that's less than two hours away."

Increasingly, SELCO is expanding beyond individual installations—the kiosk in Luckman is one—to village-wide projects. The key, Bassett says, is the "local guy who runs the kiosk." SELCO installs solar panels adjacent to the store. The electricity feeds a charging station inside the kiosk. Clipped into the station are small batteries, each the size of a cigar. At dusk, participating families send someone to fetch their battery. It connects to a SELCO 6-watt LED light via a standard VGA port (the unusual plug both helps deter theft and makes it harder to damage the devices by amateur fiddling). In the morning, the families return their battery for charging. They pay 25 rupees a month (about 40 cents) for the service. The next step, now being tested, is village solar networks—with greater capacity and independent "minigrids" that allow participants to run fans, sewing machines, and computers.

SELCO is far from alone; dozens of other solar ventures exist in the Indian countryside, though few have been as successful. Because solar energy is intermittent, many Indians see it as second class; a Greenpeace minigrid experiment in the northeastern state of Bihar last year was met by villagers chanting, "We want real electricity, not fake electricity!" But SELCO-style projects have a signal advantage: they can expand rapidly. SELCO's installations are increasing at a 20 percent annual clip. More important, the company is training 100 entrepreneurs a year to replicate its business model across the country. Instead of building huge solar parks or giant coal plants and trying to distribute electricity to remote villages, it is attempting to make the villages themselves the source of power. Hande envisions a bottom-up movement, with entrepreneurs training entrepreneurs. With luck and favorable government policies, it could represent a third path to the future—one quite different from anything as yet envisioned by Modi.

Whatever decisions India makes on the road to providing power for its hundreds of millions of unwired people, its choices will resonate around the world. Its popular prime minister has alternated between promoting renewable energy, as he did in Gujarat, and increasing the focus on coal. Neither is an easy path. Grid-style solar power requires building both massive new Charanka-style solar plants and massive energy-storage facilities, all on a scale that has never been seen in the world. It is a daunting prospect. Coal is cheaper, and there is little mystery about how to use it. But obtaining enough for India to prosper will require Coal India and other companies to sort through enormous logistical and humanitarian difficulties. And even if Modi managed to surmount them, he would be burdening India with a huge pollution problem—and the rest of the world with catastrophic carbon dioxide emissions. The nation cannot follow both paths equally. Modi, in his shifting allegiance, seems to be signaling a preference for coal.

Still, one can envision another course, in which bottom-up efforts like those from SELCO could buy some time, giving rural

Indians some of the most important benefits of electrification while allowing the nation to build up its renewables infrastructure. No serious study has yet laid out the conditions under which this could occur. But it is hard to believe this could happen without significant financial assistance from developed nations. (There is also the moral argument; as Narain said, the West did fill the atmosphere with carbon dioxide first.) India will fight hard for this in Paris. But ultimately the decision about assistance will be made by Europe and the United States.

India will make a choice, but it will not be India's alone.

EMMA MARRIS

Return of the Wild

FROM *Boom: A Journal of California*

IT IS A frosty spring morning, and I'm tracking celebrity wolves in Southern Oregon. The patriarch of this pack is a big deal. Scientists call him OR7, the seventh wolf in Oregon to be captured and fitted with a tracking collar. Environmentalists call him Journey, a name that pays homage to his epic thousand-mile trek from his birth pack in northeastern Oregon to the California border. A few days after Christmas in 2011, OR7 crossed that border, becoming the first known wild wolf in the state since 1924.[1] When OR7 found a mate—a dark black female without a collar, or a known history—they settled in Oregon, much to the disappointment of lobo fans in Los Angeles, the Bay Area, and other hotspots of California wolf fandom.

Now I think I might be looking at his poop. The sun is raising steam off the graveled timber roads. We're driving along, slowly, with a VHF receiver balanced on the front console of the truck. So far, all we've heard from the receiver is static—none of the pings that would indicate that OR7's collar is in range. But John Stephenson, the U.S. Fish and Wildlife biologist I'm accompanying, has a lead. Yesterday he followed the VHF signal to OR7 and caught a glimpse of the gray wolf and one of his offspring, but the wolves saw Stephenson first and were gone in a flash. "It is so hard to get a visual in this country," Stephenson says. "Too many damn trees."

We drive to the site of this very brief interspecies encounter and find a few monstrous piles of poop, bristling with elk hair. Stephenson bags them. We also find at least three sets of tracks headed straight down the road. Wolf prints are larger than almost all dog prints, and they typically run in these very straight lines. "It's a good way to tell a wolf from a large dog," Stephenson says. "Dogs tend to wander around."

We follow the tracks for some time, until we lose the trail on hard, dry ground. We spend the rest of the day crisscrossing the forest, seeing only the odd logging truck—no other cars or trucks, and no wolves. It's not surprising OR7 finds this nearly humanless place a good home. Wolves generally do their very best to avoid people.

With an average population density of almost 250 people per square mile, California might seem an unlikely choice for wolves in search of a home. But as far we know, wolves don't read road atlases or care about statistical averages, and there is some very wild and remote country in northeastern California—from the arid Modoc Plateau to the pine and fir forests of Mount Lassen. Stephenson wonders whether there are enough deer and elk to sustain a robust wolf population in the state; but as he prepares to document the second round of pups for this family that lives within one or two long days' walk of the California border, he says some of OR7's children could "easily" settle down in the Golden State.

Indeed, on August 20 the California Department of Fish and Wildlife announced that camera traps had caught snaps of fuzzy wolf pups playing in Northern California. They probably aren't OR7's grandchildren—Stephenson thinks they are the pups of "previously undetected dispersers from Idaho or northeast Oregon." But even if this family doesn't make a permanent home in California, the expansion of wolves into California seems inevitable. The first wolves entered neighboring Oregon in the late 1990s, the children of reintroductions undertaken by the federal government in the early 1990s in Idaho. They've found southern Oregon to be a good home, and as their numbers increase, they will almost certainly carve out additional territories in California.

The state has been preparing for their return. On June 4, 2014, the California Fish and Game Commission voted to preemptively list gray wolves as endangered under the state Endangered Species Act. The California Department of Fish and Wildlife had been completing a plan for managing the incoming wolves, though now it may need revising. Ranchers, hunters, and environmentalists have all been invited to be part of the process, and wolf advocates are feeling good about the prospects for a more cooperative, less contentious coexistence between wolves and livestock in Northern California than in the Rocky Mountains (where the "wolf wars" have turned the animal into a political football). The emphasis is on teaching the wolves not to go after livestock, by frightening them away with flagging tape, loud noises, and livestock-guarding dogs. "We are hoping to do what we can before wolves get here so it can be different," says Karin Vardaman, director of California wolf recovery at the California Wolf Center. "Because, really, if you keep politics out of it, in areas where ranchers have learned to use these nonlethal tools correctly, the controversy just went away."

The Department of Fish and Wildlife is struggling to come up with maps of where wolves could live in the state or estimates of how large a population the state potentially could host. Historical records aren't helpful—but not, as you might think, because of how much California has changed. The problem is that virtually no historical records exist. California eradicated the wolf from its landscapes so quickly and thoroughly that the animals barely appear in the historical record. It's a testament to the power of colonization and modernization that a species that was no doubt once an apex predator, one of the kings of California along with the grizzly, was reduced to a rumor, a word, a skull, a walk-on role in legend.

Indeed, until recently, it was often repeated (notably in Barry Lopez's book *Of Wolves and Men*) that there never were any wolves in California. Scientific maps showing the precontact range of wolves in North America compiled in 1944, 1953, 1981, and 2002 omit all or most of the state.[2]

Only two natural history museum specimens are verified to be California wolves from the 20th century. There are none from the 19th. "I was shocked when I started looking. How could there only be two?" says Sarah Hendricks, a geneticist who hoped to learn about the state's population dynamics by analyzing DNA from old skulls and pelts. Hendricks was working in a UCLA lab run by canine geneticist Robert Wayne when the state requested a thorough report on what was known about the vanished wolf packs' population structure just before eradication. Hendricks had only two skulls to work with, both from animals collected in the 1920s and housed in the Museum of Vertebrate Zoology at the University of

California–Berkeley. The museum sent her the tiniest sample possible—minute shavings from the inside of the precious skulls' nostrils. "I opened up the envelopes and I said, 'I don't think this is going to work, because there is hardly anything here,'" Hendricks says. But she managed to pull enough DNA from the material to establish that a wolf killed in San Bernardino County in 1922 was probably a Mexican wolf, a distinct subspecies currently being reintroduced into the wild in the Southwest. The other skull came from California's last recorded wild wolf, an emaciated, maimed critter killed in Lassen County in 1924. It had DNA markers linking it to the large population of gray wolves of the Rocky Mountains and Canada.[3] Because OR7 descends from wolves reintroduced to Idaho from inland British Columbia, Hendricks's analysis suggests that more or less the "right" kind of wolf—according to ecologists—is recolonizing Oregon and California. But with just two specimens, it is pointless to even try to guess at population densities or dynamics of these wolves.

"Other states have a frame of reference for what their populations were historically before they were eradicated," says Karen Kovacs, wildlife program manager for the California Department of Fish and Wildlife. "We scoured every source we could find." Kovacs and her team looked for trapping records. Nothing. They looked at historical accounts of the first Europeans, but she felt many of these were unreliable because of a widespread loosey-goosey habit of referring to coyotes as a kind of wolf.

Back in 1991 ecologist Robert Schmidt, then at Berkeley, combed through more than 50 European historical accounts, looking for those writers who separately mentioned and clearly distinguished between coyotes, foxes, and wolves. He also gave writers who were trained naturalists the benefit of the doubt that they knew their canids, and he found several sightings that qualified under those rules.[4] Russian explorer Otto von Kotzebue, for example, saw two species of "wolves" in the San Francisco Bay Area—most likely wolves and coyotes. In Schmidt's estimation, wolves likely lived in the Central Valley, Coast Ranges, and Sierra Nevada until about 1800. Trapping, shooting, and poisoning of these suspected livestock thieves likely occurred so quickly and so thoroughly that they were nearly lost to Euro-American history.

Of course, that's not the only history in California. Two analyses of native languages and literature have found traces of the wolf across nearly the whole state. In 2001 Alexandra Geddes-Osborne and Malcolm Margolin found separate words for "wolf" and "coyote" in many indigenous languages, and a role for wolves in story and ceremonies, in tribes as disparate as the Karuk in the far north, to the Pomo in the center of the state, and the Luiseno in the south.[5] More recently, a report by scholars from the Anthropological Studies Center at Sonoma State University found 15 indigenous languages across the state with different words for wolf, coyote, and dog, and five tribes with traditions in which the wolf features.[6]

One can go even further back in time, beyond history to prehistory. At the Page Museum at the La Brea Tar Pits in Los Angeles, little kids stare in horrified fascination at an animatronic saber-toothed cat taking down a slightly mangy-looking stuffed ground sloth in the public area. Meanwhile, drawers and drawers of specimens from Pleistocene California—from 10,000 to 50,000 years old—compose a deeper archive. Dire wolves—giant relatives of modern gray wolves, though not their ancestors—are the most common fossils found at the site. Researchers have unearthed the bones of some 4,000 individual dire wolves. Presumably, mastodons and ground sloths stuck in the tar, too, were so tempting that they lured the dire wolves to their own doom.

A curator at the tar pits looks slightly bemused that I'm less interested in the dire wolves (currently chic thanks to their appearance in *Game of Thrones*) than regular old *Canis lupus*. "Does the collection include gray wolves?" I ask.

It does, indeed. We walk down a long, narrow corridor between metal cabinets, open a drawer, and here are riches of wolf bones, looking, as tar pit specimens do, mahogany-colored and polished. There are teeth, jaws, and skulls. Nineteen drawers in all. When the first people came to what is now California, there were almost certainly wolves here.

Of course, there are still wolves in California—in Los Angeles, in fact. But these aren't free-roaming wildlife. They are pets—or prisoners, depending on your point of view. Jennifer McCarthy, a dog trainer, spent four years in Colorado studying and working with captive wolves on a large piece of land. She now applies what she learned there to the dogs of the greater Los Angeles area, including the fully domesticated, nonwolf pooches of celebrities such as Christina Aguilera, the Osborne family, and Renee Zellweger. Some of McCarthy's less famous clients own wolves or wolf-dog hybrids. For many people, there is an undeniable attraction to being that close with a piece of the wild. But wolves and wolf dogs make notoriously poor pets. They can bite. They don't follow direction well. Their predatory instincts are strong, and most of all, they are incredible escape artists. They don't bond with people the way dogs do. Wolves may sound like a cool companion animal, but they spend their lives trying to be wolves, with sometimes-disastrous consequences.

McCarthy meets me in Redondo Beach at a coffee shop, wearing a black hoodie that says WOLF WOMAN on it. "There are people who live in the city of Los Angeles with one hundred percent full-blown wolves," she says. In one case, she was called to an apartment in Beverly Hills where a wolf had chewed through the floor and escaped into the apartment below.

McCarthy disapproves of breeding and selling wolves and part-wolves. "I really believe these animals were meant to be wild," she says. "Wolves don't want a lot to do with us." But she will try to keep pet wolves from being euthanized, either by working on their behavior or placing them in one of the always-crowded specialty wolf rescues. She also volunteers her time to transport wolves and wolf dogs to shelters when necessary.

McCarthy's experience with wolves suggests to her that even if they recolonize the state in large numbers, they will stay as far away from people as possible. "I couldn't picture wolves walking down Santa Monica Boulevard at night, going through garbage cans," she says.

That's a job for another California canine: the coyote. Smaller, faster-breeding, and potentially more adaptable, the coyote inherited the lands the wolf left behind when humans exterminated it across most of the country. Coyotes, unlike wolves, are nearly impossible to eradicate. You can shoot, trap, and poison them all day long, and they'll just keep coming.

In a lot of the native California and Oregon stories in which Wolf appears, he seems to be kind of a straight man to the trickster Coyote, who is sometimes his brother. Geddes-Osborne and Margolin retell a story from the Chemehuevi,[7] in which Wolf is a brave warrior who saves the day and his little brother when the Bear people attack. Wolf fights in a magnificent multicolored robe, which becomes the rainbow. He is "wiser, more stately and in charge" than his little brother Coyote. This reminds me of stories from the Northern Paiute, recorded in 1938 by Berkeley anthropologist Isabel Kelly.[8] In these stories, Wolf and Coyote are brothers, and Coyote is constantly taking risks out of curiosity, despite the warnings of his sage, conservative older brother, Wolf. Here's a fragment of a tale told by Bige Archie of the Gidii'tikadu or Groundhog Eater band, in Modoc County, California.

> They saw someone camped. Coyote wanted to see whose camp it was. Wolf told him, "Those are pretty bad people; don't go there." Coyote thought they might have some ya'pa [camas] roots. "I'll go anyway," he said. He went over to the camp. There were some Bear women in there. There was lots of ya'pa drying outside. Coyote found a basket. He scooped up some ya'pa and ran. They came after him; they came close behind him. He threw back the basket and hit those women right on the legs. He didn't eat much of the ya'pa; he didn't have time.
>
> Coyote caught up with Wolf, and they went on.[9]

There seem to be some essential truths here about the natures of these two canines. It may be debatable whether wolves have more dignity. But they are certainly much more risk-averse. They tend to approach novel situations with the utmost caution; they shun humans; they take a long time to warm up to strange wolves. Coyotes are reckless and innovative, and as a result, humans have never managed to kill them off, in California or anywhere else. Stories about coyotes outnumber stories about wolves in most Oregon and California Indian literatures by a considerable margin. Does that mean there were fewer wolves or just that Coyote is a more compelling character for human storytellers?

Coyotes have adapted to a modern, crowded California. They cross Sunset Boulevard in San Francisco in the afternoon,[10] nibble on lychee and avocados from suburban Southern California gardens,[11] and forage in Santa Monica's exuberantly varied and rich trashcans.

I have a hard time imagining wolves in those dangerous, liminal niches. Perhaps when wolves come back to a California vastly more overrun with humans than the one they last knew, they will stay hidden in the kind of remote forests favored by OR7 and his pups —places where you can drive up and down ridges all day and hear nothing but the drone of VHF receiver static, the croaks of ravens, and the scold of nuthatches; places where you know wolves are there, but you never see them. Or perhaps wolves will surprise me and everyone else and push in close to human California, appearing on ranches, in coastal suburbs, and even in major cities.

No matter where wolves live and how many there are, humans will be watching. The leaders of California's first wolf packs likely will be caught and fitted with transmitting collars, just as in the other western states. The first colonists may well have Twitter accounts, like OR7. One thing is sure. In the first year of their official residence in the state, more will be known about them and written about them than all of the wolf generations before 1924.

Notes

1. California Department of Fish and Wildlife, "Gray Wolf *(Canis lupus)*," accessed May 11, 2015, http://dfg.ca.gov/wildlife/nongame/wolf/.
2. Stephanie L. Shelton and Floyd W. Weckerly, "Inconsistencies in Historical Geographic Range Maps: The Gray Wolf as Example," *California Fish and Game* 93, no. 4 (2007): 224.
3. Sarah A. Hendricks et al., "Polyphyletic Ancestry of Historic Gray Wolves Inhabiting US Pacific States," *Conservation Genetics* (2014): 1–6.
4. Robert H. Schmidt, "Gray Wolves in California: Their Presence and Absence," *California Fish and Game* 77, no. 2 (1991): 79–85.
5. A. Geddes-Osborne and M. Margolin, "Man and Wolf," *Defenders Magazine* 76, no. 2 (2001): 36–41.
6. M. Newland and M. Stoyka, "The Pre-Contact Distribution of Canis lupus in California: A Preliminary Assessment," unpublished paper, Sonoma State University, 2013.
7. http://www.chemehuevi.net/history-culture/.
8. http://www.nps.gov/parkhistory/online_books/joda/hrs/hrs1b.htm.
9. Isabel T. Kelly, "Northern Paiute Tales," *Journal of American Folklore* (1938): 363–438.
10. http://www.sfgate.com/outdoors/article/Coyotes-seemingly-thrive-in-San-Francisco-5045034.php.
11. http://www.ipm.ucdavis.edu/PMG/PESTNOTES/pn74135.html.

SARAH MASLIN NIR

Perfect Nails, Poisoned Workers

FROM *The New York Times*

EACH TIME A CUSTOMER pulled open the glass door at the nail shop in Ridgewood, Queens, where Nancy Otavalo worked, a cheerful chorus would ring out from where she sat with her fellow manicurists against the wall: "Pick a color!"

Ms. Otavalo, a 39-year-old Ecuadorean immigrant, was usually stationed at the first table. She trimmed and buffed and chatted about her quick-witted toddler, or her strapping 9-year-old boy. But she never spoke of another dreamed-for child, the one lost last year in a miscarriage that began while she was giving a customer a shoulder massage.

At the second table was Monica A. Rocano, 30, who sometimes brought a daughter to visit. But clients had never met her 3-year-old son, Matthew Ramon. People thought Matthew was shy, but in fact he has barely learned how to speak and can walk only with great difficulty.

A chair down from Ms. Rocano was another, quieter manicurist. In her idle moments, she surfed the Internet on her phone, seeking something that might explain the miscarriage she had last year. Or the four others that came before.

Similar stories of illness and tragedy abound at nail salons across the country, of children born slow or "special," of miscarriages and cancers, of coughs that will not go away and painful skin afflictions. The stories have become so common that older manicurists warn women of childbearing age away from the business, with its potent brew of polishes, solvents, hardeners, and glues that nail workers handle daily.

A growing body of medical research shows a link between the chemicals that make nail and beauty products useful—the ingredients that make them chip-resistant and pliable, quick to dry and brightly colored, for example—and serious health problems.

Whatever the threat the typical customer enjoying her weekly French tips might face, it is a different order of magnitude, advocates say, for manicurists who handle the chemicals and breathe their fumes for hours on end, day after day.

The prevalence of respiratory and skin ailments among nail salon workers is widely acknowledged. More uncertain, however, is their risk for direr medical issues. Some of the chemicals in nail products are known to cause cancer; others have been linked to abnormal fetal development, miscarriages, and other harm to reproductive health.

A number of studies have also found that cosmetologists—a group that includes manicurists, as well as hairdressers and makeup artists—have elevated rates of death from Hodgkin's disease, of low birth-weight babies, and of multiple myeloma, a form of cancer.

But firm conclusions are elusive, partly because the research is so limited. Very few studies have focused on nail salon workers specifically. Little is known about the true extent to which they are exposed to hazardous chemicals, what the accumulated effect is over time, and whether a connection can actually be drawn to their health.

The federal law that regulates cosmetics safety, which is more than 75 years old, does not require companies to share safety information with the Food and Drug Administration. The law bans ingredients harmful to users, but it contains no provisions for the agency to evaluate the effects of the chemicals before they are put on shelves. Industry lobbyists have fought tougher monitoring requirements.

Industry officials say their products contain minuscule amounts of the chemicals identified as potentially hazardous and pose no threat.

"What I hear are insinuations based on 'linked to,'" said Doug Schoon, cochairman of the Professional Beauty Association's Nail Manufacturers Council on Safety. "When we talk about nail polish, there's no evidence of harm."

Health advocates and officials disagree, pointing to the accumulated evidence.

"We know that a lot of the chemicals are very dangerous," said David Michaels, the assistant labor secretary who heads the federal Occupational Safety and Health Administration, which oversees workplace safety. "We don't need to see the effect in nail salon workers to know that they are dangerous to the workers."

So many health complaints were cropping up among the mostly Vietnamese manicurists in Oakland, California, that workers at Asian Health Services, a community organization there, decided on their own to investigate about a decade ago.

"It was like, 'Oh wow, what's happening in this community?'" said Julia Liou, who is now the health center's director of program planning and development and a cofounder of the California Healthy Nail Salon Collaborative. "We are seeing this epidemic of people who are sick."

The organization helped form a coalition in California that pushed for restrictions on chemicals used in nail salons, but the cosmetics industry succeeded in blocking a ban.

In recent years, in the face of growing health concerns, some polish companies have said that they have removed certain controversial chemicals from their products. But random testing of some of these products by government agencies showed the chemicals were still present.

Some states and municipalities recommend workers wear gloves and other protection, but salon owners usually discourage them from donning such unsightly gear. And even though officials overseeing workplace safety concede that federal standards on levels of chemicals that these workers can be exposed to need revision, nothing has been done.

So manicurists continue to paint fingertips, swipe off polish, and file down false nails, while absorbing chemicals that are potentially hazardous to their health.

"There are so many stories but no one that dares to tell them; no one dares to tell them because they have no one to tell," Ms. Otavalo said in an interview on a day off from the Ridgewood salon, babysitting for her colleague's developmentally disabled son, Matthew. (Ms. Otavalo left her job at the salon a few months later.) "There are thousands of women who are working in this, but no one asking: 'What's happening to you? How do you feel?' We just work and work."

"They Cannot Breathe"

The walls of Dr. Charles Hwu's second-story office in Flushing, Queens, are decorated with Chinese calligraphy, gifts from patients he has cared for from cradle to adulthood. Over his decades as an internist in this predominantly Asian enclave, Dr. Hwu has repeatedly encountered a particular set of conditions affecting otherwise healthy women.

"They come in usually with breathing problems, some symptoms similar to an allergy, and also asthma symptoms—they cannot breathe," he said during a break between patients this winter. "Judging from the symptoms with these women, it seems that they are either smokers, secondhand smokers, or asthma patients, but they are none of the above. They work for nail salons."

In interviews with over 125 nail salon workers, airway ailments like those in Dr. Hwu's office were ubiquitous. Many have learned to simply laugh them off—the nose that constantly bleeds, the throat that has ached every day since the manicurist started working.

In the nail salon she owned in Mill Basin, Brooklyn, Eugenia Colon spent years molding sometimes 30 sets of talonlike nails a day in a haze of acrylic powder, ignoring a persistent cough that grew more pronounced over time. She was found to have sarcoidosis, an inflammatory disease, in her lungs. In scans, they appeared as if covered with granules of sand, streaked by tiny scars.

The doctor who diagnosed her condition asked Ms. Colon what she did for a living. When she told him, he was frank: as she beautified other women, she inhaled clouds of acrylic and other dust, tiny particles that gouged the soft tissue of her lungs.

"We made money off it, but was it worth it?" asks Ms. Colon, 52, now an aesthetician in a Manhattan spa. "It came with a price."

Of the 20 common nail product ingredients listed as causing health problems in the appendix of a safety brochure put out by the Environmental Protection Agency, 17 are hazardous to the respiratory tract, according to the agency. Overexposure to each of them induces symptoms such as burning throat or lungs, labored breathing, or shortness of breath.

A 2006 study published in the *Journal of Occupational and Environmental Medicine* that included more than 500 Colorado manicurists found about 20 percent of them had a cough most days and nights. The same examination showed those who worked with artificial nails were about three times as likely to get asthma on the job as someone not in the industry.

Skin disorders are also omnipresent among nail salon workers. Many of the chemicals in nail salon products are classified by government agencies as skin sensitizers, capable of provoking painful reactions.

Some veteran manicurists say they can recognize one another on the street: they have the same coffee-colored stains on their cheeks. Certain cosmetic color additives—particularly a type of brilliant red—have been shown by researchers to cause such skin discoloration.

When Ki Ok Chung, a manicurist who worked in salons for almost two decades, had her fingerprints taken in the early 2000s for her United States citizenship, she made an upsetting discovery: Her prints were almost nonexistent. They had to be taken seven times. She says constant work with files, solvents, and emollients is responsible.

"I realized my fingerprints had been disappearing," she said.

Today, she cannot touch hot or cold dishes without searing pain.

Even as the weather warmed into spring last year, Zoila Calle, a manicurist, then 22, who worked in Harlem, wore wool gloves indoors and out. Underneath were black pustules so painful she could not grasp a polish bottle or text on her phone. It was the second time her hands had erupted in the warts, a common occurrence for nail salon workers. While customers often fret about salon hygiene, it is manicurists who appear truly at risk, suffering through endless fungal infections and other skin diseases from the blur of hands and feet they touch every day.

"It's a beautiful industry, it makes people feel better," Ms. Colon, who owned the salon in Mill Basin, said in an interview, a faint wheeze just audible behind her ready laugh. "But if a lot of people knew the truth behind it, it wouldn't happen. They wouldn't go."

Miscarriages and Warnings

In a way, Ms. Rocano, one of the manicurists in the Ridgewood salon, felt herself lucky. Her colleagues seated on either side of her had each lost a pregnancy last year, hoped-for babies whom they separately described exactly the same way: "Like losing a dream."

She, however, has her toddler, Matthew.

A dark-haired bundle with amber skin when he was born, Matthew was an infant laden with his mother's hopes. Holding him in her arms, she was reminded of the daughter she had left behind in Ecuador and still has not seen in more than six years. This time, she felt, she could do right by her child.

Yet as he grew, something seemed off. His legs were weak; they buckled when he tried to stand. By age three he still could not say his name. In visits to her pediatrician, she learned that Matthew was delayed on almost every measure, both physically and cognitively.

At one point, his doctor asked her what she did for a living. When she told him, he asked how long she had worked in the nail salon while pregnant. Six months, she responded.

The doctor told her, "When babies are forming in your womb, they absorb everything, and if they are exposed to anything, it can cause them harm," she recalled.

On a day five years ago, a doctor gave a similar warning to the manicurist who works at the table to Ms. Rocano's right, as she sat in the obstetrics unit at Wyckoff Heights Medical Center in Brooklyn. There she learned she had miscarried a third time.

"I went to the hospital, and I told him, 'I'm a manicurist,'" said the woman, who declined to give her name because she wanted her medical history to remain private. Her doctor urged her to change jobs. "The chemicals are not healthy for your lungs, your liver, and sometimes they begin cancer," she recalled. "I was laughing. I said, 'Who is going to pay my bills?'" She has since miscarried twice more.

In scientific circles, the three chemicals in nail products that are associated with the most serious health issues are dibutyl phthalate, toluene, and formaldehyde. They are known as the "toxic trio" among worker advocates.

Dibutyl phthalate, called DBP for short, makes nail polish and other products pliable. In Australia, it is listed as a reproductive toxicant and must be labeled with the phrases "may cause harm to the unborn child" and "possible risk of impaired fertility." Starting in June, the chemical will be prohibited from cosmetics in that country. It is one of over 1,300 chemicals banned from use in cosmetics in the European Union. But in the United States, where fewer than a dozen chemicals are prohibited in such products, there are no restrictions on DBP.

Toluene, a type of solvent, helps polish glide on smoothly. But the EPA says in a fact sheet that it can impair cognitive and kidney function. In addition, repeated exposure during pregnancy can "adversely affect the developing fetus," according to the agency.

Formaldehyde, best known for its use in embalming, is a hardening agent in nail products. In 2011 the National Toxicology Program, part of the United States Department of Health and Human Services, labeled it a human carcinogen. By 2016 it will be banned from cosmetics in the European Union.

Cosmetics industry officials say linking the chemicals to manicurists' health complaints amounts to faulty science.

Dibutyl phthalate, toluene, and formaldehyde "have been found to be safe under current conditions of use in the United States," said Lisa Powers, a spokeswoman for the Personal Care Products Council, the main trade association and lobbying group for the cosmetics industry. "The safe and historical use of these ingredients is not questioned by FDA," she continued.

In reality, the responsibility for evaluating the safety of the chemicals as they are used in cosmetics is left with the companies themselves.

Even while insisting they are safe, some polish companies have voluntarily begun to remove certain chemicals from formulations. By 2006 several prominent brands had announced their products would no longer contain any of the three. The new products were labeled "3-free" or "5-free," referring to the number of chemicals that are ostensibly no longer in them.

But a 2010 study by the FDA and another in 2012 by the California Environmental Protection Agency's Department of Toxic Substance Control found in random tests that some products, even ones labeled "3-free" or "5-free," in fact contained those very chemicals.

It was from routine community outreach trips to local nail salons in Oakland that Ms. Liou and her colleagues from Asian

Health Services, as well as Thu Quach, a research scientist, became alarmed: almost all of the manicurists interviewed had health complaints; some were terribly ill.

Dr. Quach, with the Cancer Prevention Institute of California, set out to conduct a health survey of nail salon workers in Alameda County, which includes Oakland.

The stories poured in.

Le Thi Lam, a Vietnamese immigrant who came to the United States in 1988 after fleeing the Communist government in her country, was among the first. She had started out in a Sacramento nail salon, becoming proficient in acrylic nails, sculpting them all day long from a slurry of solvent and plastic polymers.

In 1991 she learned she had a thyroid condition. She had also developed asthma. She quit, too sickened to work and concerned about the chemicals she was handling. But she soon returned, unable to find another job with her limited English. Ten years later, she had breast cancer.

"I know that manicurists like me are also going through the same things and having major health problems," she said, seated in a conference room at Asian Health Services last summer, a blouse hiding a red scar from her breastbone to her armpit. "But they still hang on to their jobs to earn their living."

Dr. Quach kept going with her research, undertaking several other studies. One found manicurists had an increased risk for gestational diabetes and for having undersize babies. Another, looking at cancer, found no correlation. Both studies were hampered by data limitations. Mostly, they point to the need for further study.

"What we know is what's reported by the women again and again: that there is something here," Dr. Quach said. "Those chemicals they are dealing with are chemicals that we know people react to, and we are hearing the stories from these workers that they are reacting to them. It's all there."

"Fox Guarding the Henhouse"

The regulation of chemicals in nail products is dictated by the Federal Food, Drug, and Cosmetic Act of 1938. The part of the law that deals with cosmetics totals just 591 words.

The Food and Drug Administration explains the limitations it faces under the law on its website: "Cosmetic products and ingredients do not need FDA premarket approval, with the exception of color additives." It continues, "Neither the law nor FDA regulations require specific tests to demonstrate the safety of individual products or ingredients." In addition, "The law also does not require cosmetic companies to share their safety information with FDA."

In 1976 the cosmetics industry itself established the Cosmetic Ingredient Review, a panel that is supposed to "review and assess the safety of ingredients used in cosmetics in an open, unbiased and expert manner," according to its website. But the panel is financed entirely by the Personal Care Products Council, the industry lobbying group. The panel's offices are also in the same building in Washington as the products council.

Even so, Ms. Powers said the panel was independent. She is the official spokeswoman for the industry lobby, but all questions to the review panel were handled by her.

Since its founding, the panel has reviewed only a small fraction of the substances in use in cosmetics today. Among them were dibutyl phthalate and toluene; the panel determined that they are safe the way they are used in nail products—on nails, not skin.

"It's a classic case of the fox guarding the henhouse," says Janet Nudelman, the director of program and policy at the Breast Cancer Fund, which has argued for more stringent regulation. "You've got an industry-funded review panel that's assessing the safety for the very industry that's funding the review panel."

There have been efforts in recent years to overhaul the 1938 law and more strictly regulate cosmetic chemicals, but none made headway in the face of industry resistance. Since 2013 the products council, just one of several industry trade groups, has poured nearly $2 million on its own into lobbying Congress.

After talks between the cosmetics industry and the FDA broke down last year, Michael R. Taylor, the agency's deputy commissioner for foods and veterinary medicine, rebuked the industry in an unusual open letter for pushing a measure that would have declared a wide range of potentially dangerous chemicals safe "without a credible scientific basis" and others safe that are known to pose "real and substantial risks to consumers."

Ms. Powers said the letter mischaracterized the industry's

stance. "The law was created or passed in 1938," she said. "Nobody is saying that we shouldn't look at that now and say: 'Is it a contemporary approach? Does it need to bring us into the 21st century?' We all agree to that. But that doesn't make for a sexy headline."

The council, in fact, said it supported a bipartisan bill introduced in April by Senators Dianne Feinstein, Democrat of California, and Susan Collins, Republican of Maine, that would broaden FDA oversight of cosmetics, including giving the agency recall ability. But some health advocates said the bill would continue to permit the industry to largely regulate itself; it would also preempt states' abilities to create stronger rules.

The Occupational Safety and Health Administration is the federal agency that sets chemical exposure limits in workplaces. The studies that have examined the chemical exposure levels for manicurists have found them to be well below these standards. Health advocates say the safety administration's standards are badly out of date and flawed.

Even Dr. Michaels, the head of the safety administration, said his agency's standards needed revision. Currently, he said, workers "can be exposed to levels that are legal according to OSHA but are still dangerous."

The agency makes illustrated pamphlets warning manicurists about the chemical hazards they face and urges them to wear gloves and ventilate their shops. These steps and others become mandatory when exposure limits are exceeded. But in practical terms, with the standards set so high, salons are free to do nothing. Dr. Michaels said the agency was hamstrung by its own cumbersome rule-making process.

"Every worker has the right to come home safely at the end of every day," Dr. Michaels said. "They shouldn't be coming home and getting sick."

The debate over the chemicals has also unfolded at the state level. In 2005 lawmakers in California proposed banning DBP from cosmetic products sold or manufactured in the state. Industry lobbyists flooded the State Capitol (some bearing gift baskets of lipstick and nail polish), spending over a half-million dollars fighting the ban, according to state records. Some of the country's best-known cosmetics companies—Estée Lauder, Mary Kay, and OPI, among others—weighed in against it. The bill ultimately failed. A much more limited measure passed—over the industry's objections—that required cosmetics companies to disclose certain hazardous chemicals to the California Department of Public Health.

Blocked by an industry with deep pockets, the California advocates say they had to scale back their goals. They introduced a grass-roots program that officially recognizes "healthy nail salons," those that carry "greener" products and that ventilate. The New York City Council held a hearing this month on a measure that would establish a similar voluntary program.

Today, out of several thousand salons in California, however, there are just 55 salons in the program.

One of them is Lulu Nail Spa, a tiny salon with a dusky rose wall and white-leather pedicure chairs in Burlingame, California. The shop earned the designation in May by switching certain products, using gloves, and opening the doors to sweep out fumes. The owner, Hai Thi Le, a Vietnamese immigrant, said she hoped the new decal she placed on her window would draw green-minded customers.

But she did not make the changes just for business. As a young woman working in her brother's nail shop, Ms. Le said she breathed in so much acrylic powder that when she kissed her husband after work, he complained her breath smelled of solvent and plastic dust.

Standing in the Breeze

On her days off from the salon in Ridgewood, Queens, Nancy Otavalo ran for a time an ad hoc daycare center at her home a few blocks away with her sister, another manicurist. The sisters would pick up salon workers' children after school for a fee, entertaining them in the basement apartment the sisters shared with their families.

Matthew, her colleague's son who can barely speak, got special treatment, spending time curled on the gleaming black leather couch—bought with tips—that is the centerpiece of her home.

After Ms. Otavalo miscarried last year, she lay for hours on the same black leather couch, in silence, the lights darkened, unable to summon the willpower to get up.

A week after a procedure to remove the fetus at Woodhull Medical Center in Brooklyn, she rose, put on the lavish makeup her sister says makes her feel confident, and went back to work at her manicure table.

Clients who stopped by for their weekly manicures knew nothing about what happened; everything appeared the same.

Except every so often, after Ms. Otavalo had painted the last stroke of top coat on a customer's hand, she scraped back her chair and walked to the front of the shop. She pulled open the salon's glass door to stand in the breeze for a while.

MADDIE OATMAN

Attack of the Killer Beetles

FROM *Mother Jones*

THERE IS AN EERIE FEEL to this grove of lodgepole pines that I can't quite put my finger on as entomologist Diana Six tromps ahead of me, hatchet in hand, scanning the southwestern Montana woods for her target. But as she digs the blade into a towering trunk, it finally hits me: the smell. There's no scent of pine needles, no sharp, minty note wafting through the brisk fall air.

Six hacks away hunks of bark until she reveals an inner layer riddled with wormy passageways. "Hey, looky!" she exclaims, poking at a small black form. "Are you dead? Yeah, you're dead." She extends her hand, holding a tiny oval maybe a quarter of an inch long. Scientists often compare this insect to a grain of rice, but Six prefers mouse dropping: "Beetle in one hand, mouse turd in another. You can't tell them apart." She turns to the next few trees in search of more traces. Pill-size holes pock their ashen trunks—a sign, along with the missing pine scent, of a forest reeling from an invasion.

These tiny winged beetles have long been culling sickly trees in North American forests. But in recent years, they've been working overtime. Prolonged droughts and shorter winters have spurred bark beetles to kill billions of trees in what's likely the largest forest insect outbreak ever recorded, about 10 times the size of past eruptions. "A doubling would have been remarkable," Six says. "Ten times screams that something is really going wrong."

Mountain pine, spruce, piñon ips, and other kinds of bark beetles have chomped 46 million of the country's 850 million acres of forested land, from the Yukon down the spine of the Rocky Mountains all the way to Mexico. Yellowstone's grizzly bears have run out of pinecones to eat because of the beetles. Skiers and backpackers have watched their brushy green playgrounds fade as trees fall down, sometimes at a rate of 100,000 trunks a day. Real estate agents have seen home prices plummet from "viewshed contamination" in areas ransacked by the bugs.

And the devastation isn't likely to let up anytime soon. As climate change warms the North American woods, we can expect these bugs to continue to proliferate and thrive in higher elevations—meaning more beetles in the coming century, preying on bigger chunks of the country.

In hopes of staving off complete catastrophe, the United States Forest Service, which oversees 80 percent of the country's woodlands, has launched a beetle offensive, chopping down trees to prevent future infestations. The USFS believes this strategy reduces trees' competition for resources, allowing the few that remain to better resist invading bugs. This theory just so happens to also benefit loggers, who are more than willing to help thin the forests. Politicians, too, have jumped onboard, often on behalf of the timber industry: more than 50 bills introduced since 2001 in Congress proposed increasing timber harvests in part to help deal with beetle outbreaks.

But Six believes that the blitz on the bugs could backfire in a big way. For starters, she says, cutting trees "quite often removes more trees than the beetles would"—effectively outbeetling the beetles. But more importantly, intriguing evidence suggests that the bugs might be on the forest's side. Six and other scientists are beginning to wonder: What if the insects that have wrought this devastation actually know more than we do about adapting to a changing climate?

Though they're often described as pesky invaders, bark beetles have been a key part of conifer ecosystems for ages, ensuring that groves don't get overcrowded. When a female mountain pine beetle locates a frail tree, she emits a chemical signal to her friends, who swarm to her by the hundreds. Together they chew through the bark until they reach the phloem, a cushy resinous layer between the outer bark and the sapwood that carries sugars through the tree. There, they lay their eggs in tunnels, and eventually a new generation of beetles hatches, grows up, and flies away. But before they do, the mature beetles also spread a special fungus in the center of the trunk. And that's where things get really interesting.

Six focuses on the "evolutionary marriage" of beetle and fungi at her four-person lab at the University of Montana, where she is the chair of the Department of Ecosystem and Conservation Sciences. Structures in bark beetles' mouths have evolved to carry certain types of fungi that convert the tree's tissue into nutrients for the bug. The fungi have "figured out how to hail the beetle that will get them to the center of the tree," Six says. "It's like getting a taxi." The fungi leave blue-gray streaks in the trees they kill; "blue-stain pine" has become a specialty product, used to make everything from cabins to coffins to iPod cases.

A healthy tree can usually beat back invading beetles by deploying chemical defenses and flooding them out with sticky resin. But just as dehydration makes humans weaker, heat and drought impede a tree's ability to fight back—less water means less resin. In some areas of the Rocky Mountain West, the mid-2000s was the driest, hottest stretch in 800 years. From 2000 to 2012 bark beetles killed enough trees to cover the entire state of Colorado. "Insects reflect their environment," explains renowned entomologist Ken Raffa—they serve as a barometer of vast changes taking place in an ecosystem.

Typically, beetle swells subside when they either run out of trees or when long, cold winters freeze them off (though some larvae typically survive, since they produce antifreeze that can keep them safe down to 30 below). But in warm weather the bugs thrive. In 2008 a team of biologists at the University of Colorado observed pine beetles flying and attacking trees in June, a month earlier than previously recorded. With warmer springs, the beetle flight season had doubled, meaning they could mature and lay eggs—and then their babies could mature and lay eggs—all within one summer.

That's not the only big change. Even as the mountain pine beetles run out of lodgepole pines to devour in the United States, in 2011 the insects made their first jump into a new species of tree, the jack pine, in Alberta. "Those trees don't have evolved defenses," Six says, "and they're not fighting back." The ability to invade a new species means the insects could begin a trek east across Canada's boreal forest, then head south into the jack, red, and white pines of Minnesota and the Great Lakes region, and on to the woods of the East Coast. Similarly, last year, the reddish-black spruce beetle infested five times as many acres in Colorado as it did in 2009. And in the last decade, scientists spotted the southern pine beetle north of the Mason-Dixon Line for the first time on record, in New Jersey and later on Long Island. As investigative journalist Andrew Nikiforuk put it in his 2011 book on the outbreaks, we now belong to the "empire of the beetle."

In a weird way, all of this is exciting news for Six: She is not only one of the world's foremost experts in beetle-fungi symbiosis, but proud to be "one of the few people in Montana that thinks bark beetles are cute." (She's even brewed her own beer from beetle fungi.) As a child, she filled her bedroom in Upland, California, with jars of insects and her fungus collection. But as a teenager, she got into drugs, quit high school, and started living on the streets. Nine years later, she attended night school, where teachers urged her to become the first in her family to go to college. And when she finally did, she couldn't get enough: classes in microbiology and integrated pest management led to a master's degree in veterinary entomology, then a PhD in entomology and mycology and a postdoc in chemical ecology, focused on insect pheromones.

Six, 58, has light-green eyes ringed with saffron, and long silvery-blond hair streaming down shoulders toned from fly-fishing and bodybuilding. As several fellow researchers stress to me, she is the rare scientist who's also a powerful communicator. "I think about what it means to be a tree," she told a rapt audience at a TEDx talk about global forest die-offs. "Trees can't walk. Trees can't run. Trees can't hide," she continued, her sonorous voice pausing carefully for emphasis. "And that means, when an enemy like the mountain pine beetle shows up, they have no choice but to stand their ground."

To a tree hugger, that might seem a grim prognosis: Since trees can't escape, they'll all eventually be devoured by insects, until we have no forests left. Especially since, with our current climate projections, we might be headed toward a world in which beetle blooms do not subside easily and instead continue to spread through new terrain.

But Six has a different way of looking at the trees' plight: as a battle for survival, with the army of beetles as a helper. She found compelling evidence of this after stumbling across the work of

Forest Service researcher Constance Millar, with whom she had crossed paths at beetle conferences.

Millar was comparing tree-core measurements of limber pines, a slight species found in the Eastern Sierras of California that can live to be 1,000 years old. After mountain pine beetles ravaged one of her study sites in the late 1980s, certain trees survived. They were all around the same size and age as the surrounding trees that the beetles tore through, so Millar looked closer at tree-ring records and began to suspect that, though they looked identical on the outside, the stand in fact had contained two genetically distinct groups of trees. One group had fared well during the 1800s, when the globe was still in the Little Ice Age and average temperatures were cooler. But this group weakened during the warmer 1900s, and grew more slowly as a result. Meanwhile, the second group seemed better suited for the warmer climate, and started to grow faster.

When beetle populations exploded in the 1980s, this second group mounted a much more successful battle against the bugs. After surviving the epidemic, this group of trees "ratcheted forward rapidly," Millar explains. When an outbreak flared up in the mid-2000s, the bugs failed to infiltrate any of the survivor trees in the stand. The beetles had helped pare down the trees that had adapted to the Little Ice Age, leaving behind the ones better suited to hotter weather. Millar found similar patterns in whitebark pines and thinks it's possible that this type of beetle-assisted natural selection is going on in different types of trees all over the country.

When Six read Millar's studies, she was floored. Was it possible, she wondered, that we've been going about beetle management all wrong? "It just hit me," she says. "There is something amazing happening here."

Last year Six and Eric Biber, a University of California–Berkeley law professor, published a provocative review paper in the journal *Forests* that challenged the Forest Service's beetle-busting strategies. After scrutinizing every study about beetle control that they could get their hands on, they concluded that "even after millions of dollars and massive efforts, suppression . . . has never effectively been achieved, and, at best, the rate of mortality of trees was reduced only marginally."

Six points to a stand of lodgepoles in the University of Montana's Lubrecht Experimental Forest. In the early 2000s school foresters preened the trees, spacing them out at even distances, and hung signs to note how this would prevent beetle outbreaks. This "prethinned" block was "the pride and joy of the experimental forest," Six remembers. But that stand was the first to get hit by encroaching pine beetles, which took out every last tree. She approached the university forest managers. "I said, 'Boy, you need to document that,'" Six says. "They didn't. They just cut it down. Now there's just a field of stumps."

Six and Biber's paper came as a direct affront to some Forest Service researchers, one of whom told me that he believes changing forest structure through thinning is the only long-term solution to the beetle problem. Politicians tend to agree—and beetle suppression sometimes serves as a convenient excuse: "It is perhaps no accident that the beetle treatments most aggressively pushed for in the political landscape allow for logging activities that provide revenue and jobs for the commercial timber industry," Six and Biber wrote in the *Forests* review.

Take the Restoring Healthy Forests for Healthy Communities Act, proposed in 2013 by then representive Doc Hastings (R-Wash.) and championed by then representative Steve Daines (R-Mont.). The bill sought to designate "revenue areas" in every national forest where, to help address insect infestations, loggers would be required to clear a certain number of trees every year. Loggers could gain access to roadless areas, wilderness-study areas, and other conservation sites, and once designated, their acreage could never be reduced. The zones would also be excluded from the standard environmental-review process.

Six and other scientists vehemently opposed these massive timber harvests—as did environmental advocates like the Sierra Club and Defenders of Wildlife, the latter warning that the harvests would take logging to "unprecedented and unsustainable levels." The bill passed the House but died in the Senate last year. But Daines, now a senator and one of 2014's top-10 recipients of timber money, vows to renew the effort so as to "revitalize Montana's timber industry" and "protect the environment for future generations."

This summer, Six plans to start examining the genes of "supertrees" —those that survive beetle onslaughts—in stands of whitebarks in

Montana's Big Hole Valley. Her findings could help inform a new kind of forest management guided by a deeper understanding of tree genes—one that beetles have had for millennia.

If we pay close enough attention, someday we may be able to learn how to think like they do. University of California–Davis plant sciences professor David Neale champions a new discipline called "landscape genomics." At his lab in Davis, Neale operates a machine that grinds up a tree's needles and spits out its DNA code. This technology is already being used for fruit-tree breeding and planting, but Neale says it could one day be used in wild forests. "As a person, you can take your DNA and have it analyzed, and they can tell you your relative risk to some disease," Neale says. "I'm proposing to do the same thing with a tree: I can estimate the relative risk to a change in temperature, change in moisture, introduction to a pathogen."

Right now, foresters prune woodlands based on the size of trees' trunks and density of their stands. If we knew more about trees' genetic differences, Neale says, "maybe we would thin the ones that have the highest relative risks." This application is still years off, but Neale has already assembled a group of Forest Service officials who want to learn more about landscape genomics.

Six, meanwhile, places her faith in the beetles. Whereas traditional foresters worry that failing to step in now could destroy America's forests, Six points to nature's resilience. Asked at TEDx how she wants to change the world, she responded, "I don't want to change the world. We have changed the world to a point that it is barely recognizable. I think it's time to stop thinking change and try to hold on to what beauty and function remains."

STEPHEN ORNES

The Whole Universe Catalog

FROM *Scientific American*

A SEEMINGLY ENDLESS VARIETY of food was sprawled over several tables at the home of Judith L. Baxter and her husband, mathematician Stephen D. Smith, in Oak Park, Illinois, on a cool Friday evening in September 2011. Canapés, homemade meatballs, cheese plates, and grilled shrimp on skewers crowded against pastries, pâtés, olives, salmon with dill sprigs, and feta wrapped in eggplant. Dessert choices included—but were not limited to—a lemon mascarpone cake and an African pumpkin cake. The sun set, and champagne flowed, as the 60 guests, about half of them mathematicians, ate and drank and ate some more.

The colossal spread was fitting for a party celebrating a mammoth achievement. Four mathematicians at the dinner—Smith, Michael Aschbacher, Richard Lyons, and Ronald Solomon—had just published a book, more than 180 years in the making, that gave a broad overview of the biggest division problem in mathematics history.

Their treatise did not land on any bestseller lists, which was understandable, given its title: *The Classification of Finite Simple Groups.* But for algebraists, the 350-page tome was a milestone. It was the short version, the CliffsNotes, of this universal classification. The full proof reaches some 15,000 pages—some say it is closer to 10,000—that are scattered across hundreds of journal articles by more than 100 authors. The assertion that it supports is known, appropriately, as the Enormous Theorem. (The theorem itself is quite simple. It is the proof that gets gigantic.) The cornucopia at Smith's house seemed an appropriate way to honor this behemoth. The proof is the largest in the history of mathematics.

And now it is in peril. The 2011 work sketches only an outline of the proof. The unmatched heft of the actual documentation places it on the teetering edge of human unmanageability. "I don't know that anyone has read everything," says Solomon, age 66, who studied the proof his entire career. (He retired from Ohio State University two years ago.) Solomon and the other three mathematicians honored at the party may be the only people alive today who understand the proof, and their advancing years have everyone worried. Smith is 67, Aschbacher is 71, and Lyons is 70. "We're all getting old now, and we want to get these ideas down before it's too late," Smith says. "We could die, or we could retire, or we could forget."

That loss would be, well, enormous. In a nutshell, the work brings order to group theory, which is the mathematical study of symmetry. Research on symmetry, in turn, is critical to scientific areas such as modern particle physics. The Standard Model—the cornerstone theory that lays out all known particles in existence, found and yet to be found—depends on the tools of symmetry provided by group theory. Big ideas about symmetry at the smallest scales helped physicists figure out the equations used in experiments that would reveal exotic fundamental particles, such as the quarks that combine to make the more familiar protons and neutrons.

Group theory also led physicists to the unsettling idea that mass itself—the amount of matter in an object such as this magazine, you, everything you can hold and see—formed because symmetry broke down at some fundamental level. Moreover, that idea pointed the way to the discovery of the most celebrated particle in recent years, the Higgs boson, which can exist only if symmetry falters at the quantum scale. The notion of the Higgs popped out of group theory in the 1960s but was not discovered until 2012, after experiments at CERN's Large Hadron Collider near Geneva.

Symmetry is the concept that something can undergo a series of transformations—spinning, folding, reflecting, moving through time—and, at the end of all those changes, appear unchanged. It lurks everywhere in the universe, from the configuration of quarks to the arrangement of galaxies in the cosmos.

The Enormous Theorem demonstrates with mathematical precision that any kind of symmetry can be broken down and grouped into one of four families, according to shared features. For mathematicians devoted to the rigorous study of symmetry, or group theorists, the theorem is an accomplishment no less sweeping, important, or fundamental than the periodic table of the elements was for chemists. In the future, it could lead to other profound discoveries about the fabric of the universe and the nature of reality.

Except, of course, that it is a mess: the equations, corollaries, and conjectures of the proof have been tossed amid more than 500 journal articles, some buried in thick volumes, filled with the mixture of Greek, Latin, and other characters used in the dense language of mathematics. Add to that chaos the fact that each contributor wrote in his or her idiosyncratic style.

That mess is a problem because without every piece of the proof in position, the entirety trembles. For comparison, imagine the two-million-plus stones of the Great Pyramid of Giza strewn haphazardly across the Sahara, with only a few people who know how they fit together. Without an accessible proof of the Enormous Theorem, future mathematicians would have two perilous choices: simply trust the proof without knowing much about how it works or reinvent the wheel. (No mathematician would ever be comfortable with the first option, and the second option would be nearly impossible.)

The 2011 outline put together by Smith, Solomon, Aschbacher, and Lyons was part of an ambitious survival plan to make the theorem accessible to the next generation of mathematicians. "To some extent, most people these days treat the theorem like a black box," Solomon laments. The bulk of that plan calls for a streamlined proof that brings all the disparate pieces of the theorem together. The plan was conceived more than 30 years ago and is now only half-finished.

If a theorem is important, its proof is doubly so. A proof establishes the honest dependability of a theorem and allows one mathematician to convince another—even when separated by continents or centuries—of the truth of a statement. Then these statements beget new conjectures and proofs, such that the collaborative heart of mathematics stretches back millennia.

Inna Capdeboscq of the University of Warwick in England is

one of the few younger researchers to have delved into the theorem. At age 44, soft-spoken and confident, she lights up when she describes the importance of truly understanding how the Enormous Theorem works. "What is classification? What does it mean to give you a list?" she ponders. "Do we know what every object on this list is? Otherwise, it's just a bunch of symbols."

Reality's Deepest Secrets

Mathematicians first began dreaming of the proof at least as early as the 1890s, as a new field called group theory took hold. In math, the word "group" refers to a set of objects connected to one another by some mathematical operation. If you apply that operation to any member of the group, the result is yet another member.

Symmetries, or movements that do not change the look of an object, fit this bill. Consider, as an example, that you have a cube with every side painted the same color. Spin the cube 90 degrees —or 180 or 270—and the cube will look exactly as it did when you started. Flip it over, top to bottom, and it will appear unchanged. Leave the room and let a friend spin or flip the cube—or execute some combination of spins and flips—and when you return, you will not know what he or she has done. In all, there are 24 distinct rotations that leave a cube appearing unchanged. Those 24 rotations make a *finite* group.

Simple finite groups are analogous to atoms. They are the basic units of construction for other, larger things. Simple finite groups combine to form larger, more complicated finite groups. The Enormous Theorem organizes these groups the way the periodic table organizes the elements. It says that every simple finite group belongs to one of three families—or to a fourth family of wild outliers. The largest of these rogues, called the Monster, has more than 10^{53} elements and exists in 196,883 dimensions. (There is even a whole field of investigation called monsterology in which researchers search for signs of the beast in other areas of math and science.) The first finite simple groups were identified by 1830, and by the 1890s mathematicians had made new inroads into finding more of those building blocks. Theorists also began to suspect the groups could all be put together in a big list.

Mathematicians in the early 20th century laid the foundation for the Enormous Theorem, but the guts of the proof did not materialize until midcentury. Between 1950 and 1980—a period which mathematician Daniel Gorenstein of Rutgers University called the "Thirty Years' War"—heavyweights pushed the field of group theory further than ever before, finding finite simple groups and grouping them together into families. These mathematicians wielded 200-page manuscripts like algebraic machetes, cutting away abstract weeds to reveal the deepest foundations of symmetry. (Freeman Dyson of the Institute for Advanced Study in Princeton, New Jersey, referred to the onslaught of discovery of strange, beautiful groups as a "magnificent zoo.")

Those were heady times: Richard Foote, then a graduate student at the University of Cambridge and now a professor at the University of Vermont, once sat in a dank office and witnessed two famous theorists—John Thompson, now at the University of Florida, and John Conway, now at Princeton University—hashing out the details of a particularly unwieldy group. "It was amazing, like two Titans with lightning going between their brains," Foote says. "They never seemed to be at a loss for some absolutely wonderful and totally off-the-wall techniques for doing something. It was breathtaking."

It was during these decades that two of the proof's biggest milestones occurred. In 1963 a theorem by mathematicians Walter Feit and John Thompson laid out a recipe for finding more simple finite groups. After that breakthrough, in 1972 Gorenstein laid out a 16-step plan for proving the Enormous Theorem—a project that would, once and for all, put all the finite simple groups in their place. It involved bringing together all the known finite simple groups, finding the missing ones, putting all the pieces into appropriate categories, and proving there could not be any others. It was big, ambitious, unruly, and, some said, implausible.

The Man with the Plan

Yet Gorenstein was a charismatic algebraist, and his vision energized a new group of mathematicians—with ambitions neither simple nor finite—who were eager to make their mark. "He was a larger-than-life personality," says Lyons, who is at Rutgers. "He was tremendously aggressive in the way he conceived of problems and conceived of solutions. And he was very persuasive in convincing other people to help him."

Solomon, who describes his first encounter with group theory as "love at first sight," met Gorenstein in 1970. The National Science Foundation was hosting a summer institute on group theory at Bowdoin College, and every week mathematical celebrities were invited to the campus to give a lecture. Solomon, who was then a graduate student, remembers Gorenstein's visit vividly. The mathematical celebrity, just arrived from his summer home on Martha's Vineyard, was electrifying in both appearance and message.

"I'd never seen a mathematician in hot-pink pants before," Solomon recalls.

In 1972, Solomon says, most mathematicians thought that the proof would not be done by the end of the 20th century. But within four years the end was in sight. Gorenstein largely credited the inspired methods and feverish pace of Aschbacher, who is a professor at the California Institute of Technology, for hastening the proof's completion.

One reason the proof is so huge is that it stipulates that its list of finite simple groups is complete. That means the list includes every building block, and there are not any more. Oftentimes proving something does not exist—such as proving there cannot be any more groups—is more work than proving it does.

In 1981 Gorenstein declared the first version of the proof finished, but his celebration was premature. A problem emerged with a particularly thorny 800-page chunk, and it took some debate to resolve it successfully. Mathematicians occasionally claimed to find other flaws in the proof or to have found new groups that broke the rules. To date, those claims have failed to topple the proof, and Solomon says he is fairly confident that it will stand.

Gorenstein soon saw the theorem's documentation for the sprawling, disorganized tangle that it had become. It was the product of a haphazard evolution. So he persuaded Lyons—and in 1982 the two of them ambushed Solomon—to help forge a revision, a more accessible and organized presentation, which would become the so-called second-generation proof. Their goals were to lay out its logic and keep future generations from having to reinvent the arguments, Lyons says. In addition, the effort would whittle the proof's 15,000 pages down, reducing it to a mere 3,000 or 4,000.

Gorenstein envisioned a series of books that would neatly collect all the disparate pieces and streamline the logic to iron over idiosyncrasies and eliminate redundancies. In the 1980s the proof was inaccessible to all but the seasoned veterans of its forging. Mathematicians had labored on it for decades, after all, and wanted to be able to share their work with future generations. A second-generation proof would give Gorenstein a way to assuage his worries that their efforts would be lost amid heavy books in dusty libraries.

Gorenstein did not live to see the last piece put in place, much less raise a glass at the Smith and Baxter house. He died of lung cancer on Martha's Vineyard in 1992. "He never stopped working," Lyons recalls. "We had three conversations the day before he died, all about the proof. There were no goodbyes or anything; it was all business."

Proving It Again

The first volume of the second-generation proof appeared in 1994. It was more expository than a standard math text and included only 2 of 30 proposed sections that could entirely span the Enormous Theorem. The second volume was published in 1996, and subsequent ones have continued to the present—the sixth appeared in 2005.

Foote says the second-generation pieces fit together better than the original chunks. "The parts that have appeared are more coherently written and much better organized," he says. "From a historical perspective, it's important to have the proof in one place. Otherwise, it becomes sort of folklore, in a sense. Even if you believe it's been done, it becomes impossible to check."

Solomon and Lyons are finishing the seventh book this summer, and a small band of mathematicians have already made inroads into the eighth and ninth. Solomon estimates that the streamlined proof will eventually take up 10 or 11 volumes, which means that just more than half of the revised proof has been published.

Solomon notes that the 10 or 11 volumes *still* will not entirely cover the second-generation proof. Even the new, streamlined version includes references to supplementary volumes and previous theorems, proved elsewhere. In some ways, that reach speaks to the cumulative nature of mathematics: every proof is a product not only of its time but of all the thousands of years of thought that came before.

In a 2005 article in the *Notices of the American Mathematical Society,* mathematician E. Brian Davies of King's College London pointed out that the "proof has never been written down in its entirety, may never be written down, and as presently envisaged would not be comprehensible to any single individual." His article brought up the uncomfortable idea that some mathematical efforts may be too complex to be understood by mere mortals. Davies's words drove Smith and his three coauthors to put together the comparatively concise book that was celebrated at the party in Oak Park.

The Enormous Theorem's proof may be beyond the scope of most mathematicians—to say nothing of curious amateurs—but its organizing principle provides a valuable tool for the future. Mathematicians have a long-standing habit of proving abstract truths decades, if not centuries, before they become useful outside the field.

"One thing that makes the future exciting is that it is difficult to predict," Solomon observes. "Geniuses come along with ideas that nobody of our generation has had. There is this temptation, this wish and dream, that there is some deeper understanding still out there."

The Next Generation

These decades of deep thinking did not only move the proof forward; they built a community. Judith Baxter—who trained as a mathematician—says group theorists form an unusually social group. "The people in group theory are often lifelong friends," she observes. "You see them at meetings, travel with them, go to parties with them, and it really is a wonderful community."

Not surprisingly, these mathematicians who lived through the excitement of finishing the first iteration of the proof are eager to preserve its ideas. Accordingly, Solomon and Lyons have recruited other mathematicians to help them finish the new version and preserve it for the future. That is not easy: many younger mathematicians see the proof as something that has already been done, and they are eager for something different.

In addition, working on rewriting a proof that has already been established takes a kind of reckless enthusiasm for group theory. Solomon found a familiar devotee to the field in Capdeboscq, one of a handful of younger mathematicians carrying the torch for the completion of the second-generation proof. She became enamored of group theory after taking a class from Solomon.

"To my surprise, I remember reading and doing the exercises and thinking that I loved it. It was beautiful," Capdeboscq says. She got "hooked" on working on the second-generation proof after Solomon asked for her help in figuring out some of the missing pieces that would eventually become part of the sixth volume. Streamlining the proof, she says, lets mathematicians look for more straightforward approaches to difficult problems.

Capdeboscq likens the effort to refining a rough draft. Gorenstein, Lyons, and Solomon laid out the plan, but she says it is her job, and the job of a few other youngsters, to see all the pieces fall into place: "We have the road map, and if we follow it, at the end the proof should come out."

RINKU PATEL

Bugged

FROM *Popular Science*

SOMEONE ONCE TOLD ME that a praying mantis in your home brings luck and good health. As for the one sitting on my kitchen countertop in Oakland, California, well, Jonathan Eisen certainly likes it. "That's cool," says the University of California at Davis microbiologist, lifting the tiny aluminum toy—with huge eyes and delicate clawlike front legs—off the cold marble. He sets it down only when something even smaller, a fruit fly, buzzes past. "Look," he says admiringly, head cocked to my ceiling, "you have drosophila."

Eisen is a tall guy in his 40s with a mountain-man beard, and he has shown up at my home wearing a T-shirt with sparkly pink block lettering that reads: ASK ME ABOUT FECAL TRANSPLANTS. He's a firm believer that human health depends on bugs—not the six-legged variety, but the microbes that populate our guts and the environments in which we live, work, and play. Eisen explains that every time I open my door, a blast of air that has woven through the surrounding tree canopy carries microbes into my house—as do Amazon packages, pets, and muddy feet.

He's musing about my oak trees when the forced-air heating clicks on. The furrows in his brow deepen. Hot, dry air shooting through a sealed house kills germs, he tells me. In fact, my whole house makes him deeply uncomfortable. It was extensively remodeled this past summer with antimicrobial fixtures, floors, and walls—now standard in many renovations. Eisen compares this practice to the overuse of antibiotics in medicine: wipe out the natural balance of good bugs, and you might not like the organisms that survive.

A mounting body of research has shown the importance of the microbes that live inside us, and scientists have been slowly cataloging species that live outside in nature. But little is known about the microbial ecosystem that surrounds us indoors, where we spend about 90 percent of our time. Recently a group of scientists, loosely connected through the Microbiology of the Built Environment Network that Eisen founded, has begun to probe it. The White House Office of Science and Technology Policy is looking into forming a national initiative to spur further research. Once we know what organisms we live with, we can begin to determine how we rely on them—and then we can tackle this question: To what extent do we need to stop protecting people from germs and instead protect germs from people?

I lead Eisen up a stairwell slathered in antimicrobial paint, and into a study with carpet treated with stain and odor guard. "You know that's bad, right?" he asks. Then we pop into the bathroom. Eisen stares intensely at the tankless toilet. It appears to levitate off the floor like an antimicrobial spaceship. When I ask if he wants to step outside for fresh air, he looks relieved.

Charles Darwin, in *On the Origin of Species,* charts evolution through the Tree of Life. Its branches and roots lift some species toward fecundity while knocking others down to extinction. But Darwin's tree didn't include microbes, perhaps the most successful life forms of all. They make up roughly 60 percent of Earth's biomass. There are more microbes in a teaspoon of soil than there are humans in the world.

By some measures, even we are more microbe than mammal. The trillions of microorganisms we harbor in our bodies, collectively known as our microbiome, outnumber human cells 10 to 1. Altogether, they weigh up to twice as much as the human brain, existing as a sort of sixth human superorgan whose function is linked to digesting our meals, preventing infection, and possibly even influencing our emotions and moods. Studies that describe new and essential roles for our microbiome are published almost daily. The reason for its breathtaking range is simple: our germs have evolved with us.

Microbes appear to have prospered by making themselves incredibly useful, and we've gladly given up space in exchange for the vitamins, digestive enzymes, and metabolites they provide. And so the discovery that the urban gut harbors up to 40 percent less microbial diversity than that of indigenous people living in a remote jungle concerns scientists. These "missing microbes," they say, may have been decimated by several decades of industrialized foods, which limited our diets, and antibiotic use, which extended our lives at the expense of theirs.

Eisen offers another explanation for why our internal real estate might be in subprime condition: the microbiome within us depends upon the microbiome that surrounds us. "Have you seen germ-free mice?" he asks me. "They are seriously messed-up animals." Delivered by cesarean section and raised in sterile chambers, these rodents have inflamed lungs and colons, like those seen in asthma and colitis. They're also prone to haywire immunity and weird social tics.

Until relatively recently, sterile chambers weren't our environments either. "We didn't evolve in closed rooms," says Maria Gloria Dominguez-Bello, a microbiologist at New York University who led the indigenous-microbiome study. "We evolved in nature." Big families lived together on farms and in tenements, not exactly temples of hygiene. Livestock loped in the streets. Infectious disease rippled through cities. Roofs leaked. Sewers overflowed. Windows opened. But with modernization, we sealed ourselves away. In other words, we parted ways with the microbes that evolved with us. By redesigning our buildings, we redesigned ourselves.

Soft of heart and loud of mouth, Eisen enjoys a good jab. When I first met him at a Thai restaurant in Davis, he lifted up his shirt and stabbed himself with an insulin syringe. I flinched, but he grinned. "When I was a kid, I did this to freak people out," he said. Now, he's illustrating how his work in the field of microbiology is personal. Eisen has type 1 diabetes, an autoimmune disease linked to, among other things, changes in the microbiome.

To understand how seriously Eisen takes his position as the defender of microbial diversity, it's useful to know where he got his career start: in an undergraduate internship at the D.C. Public Defender Service. It fostered a lifelong ardor for justice and an impulse to, whenever possible, stick it to the bullies. He argues that microbial communities—whether in our bodies or in buildings

—function as complex ecosystems, not unlike tropical rainforests. "That doesn't mean microbes don't kill some people and make others sick," he says. "But if you're afraid of a tiger, you don't clear-cut the rainforest. Well, you do in some cases, but that's crazy."

Until last year, Eisen was a member of the Forum on Microbial Threats. (He quit, saying both beneficial microbes and female scientists were underrepresented.) At the time the National Academy of Sciences first convened the forum, the prevailing narrative was that microbes were an enemy of public health and we were at war with them. The approach backfired: germs adapt to whatever drugs are thrown at them, swapping genes with neighbors to accrue antibiotic resistance. The rise of superbugs, coupled with growing awareness of the human microbiome, has led many scientists, including the forum, to rethink the merits of germ warfare.

Eisen takes a bite of stir-fry and suggests we ditch the word "pathogen" altogether. "Sometimes germs are good, sometimes they're bad," he says, sounding unusually Yoda-like. "Nothing is good or bad all the time."

As someone who has spent 20 years studying microbial evolution, Eisen is in a good position to explain the paradigm shift. In 2007 he helped launch a "genomic encyclopedia" of microbes—a splashy debut whose biggest point was all of the blank pages: we have no idea who the vast majority of our microbial neighbors are.

That hasn't stopped us from trying to kick them out. There are now thousands of antimicrobial products on the market, which range from clothing to cutting boards. One industry report forecasts that the $1.9 billion coating market alone will more than double in 2020. Rolf Halden, an Arizona State University environmental engineer, says the marketing preys on consumers' fears. "There's ample evidence we use too many antimicrobials," he says, "and without judgment."

Halden has found that triclosan, a common antimicrobial, makes its way from products like hand soap into sewage, where it breeds antibiotic resistance. Studies have also detected high levels of triclosan in house dust. One found it counterintuitively helps staphylococcus—a common source of infection—adhere to plastic and glass surfaces. What we don't know is how it or other antimicrobials affect the organisms that might actually help us.

This topic makes Eisen visibly agitated. He waves his fist like a trial lawyer itching to clock opposing counsel. He brings up a company hawking a new indoor sanitation technology on Twitter—a 24-hour, Purell-like system that purportedly kills everything, including Ebola. It's an indiscrimate weapon in the old war. Struggling for composure, he says: "*That* doesn't sound good."

In order to understand what happens when a built environment's microbial ecosystem is wiped out, scientists have begun to study the most sterile structures on Earth—and off. For astronauts, the International Space Station (ISS) is like living inside a giant antibiotic pill. HEPA filters remove airborne germs, surfaces deter bacterial growth, and iodine and biocidal nanosilver cull microbes from water. "Everything is sterilized, except for the humans," says Hernan Lorenzi of the J. Craig Venter Institute, which has been studying the ISS for four years.

As a result, the microbial ecosystem in the station is made up mostly of the organisms the astronauts themselves shed daily. There are no Amazon deliveries, no windows to crack—no influx of fresh microbes to balance the ecosystem. And so Lorenzi's team is sampling the microbiome of astronauts to see how it changes after a stint in the station. A loss of gut diversity, he says, correlates with many diseases and could raise concerns for long-term space travel. Astronauts often have impaired immunity, and "if you lose your gut microflora," Lorenzi says, "the immune system goes dormant." It takes a space vacation. "Can you imagine a trip to Mars?" asks Eisen. "They've gotta be screwed."

On Earth, the same phenomenon occurs in hospitals, only sick patients are the ones shedding microbes. Despite extensive sanitation, infections acquired in U.S. hospitals kill about 75,000 people annually—more deaths than from breast cancer and HIV/AIDS combined. The Chicago-based Hospital Microbiome Project, led by Argonne National Laboratory's Jack Gilbert, studied the ecology of one hospital for a year and found microbes everywhere. "You can do as much cleaning as you want," says Gilbert. "The hospital is a bloody sterile place, and a pathogen might still make you sick."

That sounds terrifying, but everyone harbors pathogens. The dreaded *Clostridium difficile*, which can cause life-threatening diarrhea, is found in 66 percent of infants. *Staphylococcus aureus* is carried by 20 percent of adults. People who seem perfectly healthy harbor the influenza virus. These germs don't do much harm when they're kept in check by other organisms. Studies suggest, for instance, that the flu virus can be contained through competition with *Lactobacillus.*

And so Gilbert thinks the notion that we "catch" things is flawed. In a study of intensive care units, his group observed otherwise harmless microbes go rogue in four patients after drugs decimated their gut flora. "You put humans through the ringer, and we're surprised their germs are stressed too?" he asks. Scientists suspect that in hospital rooms, sanitization can likewise pressure microbes to evolve into virulent pathogens, which then colonize surfaces cleared of competitive bacteria. Recycled-air systems help concentrate them. "We've gone too far," says Gilbert. "Hygiene is good; sterility may not be."

For Sandra Bauder, an architect in Houston, the zealous sanitation trend brings to mind a fancy horse her uncle kept in Venezuela. "He babied it—with special food, an air-conditioned barn, never let any bugs get on it. And it was always sick. Then he got a mutt horse. It lived in a pasture. It didn't get anything, not even a stomachache. I think it's the same for people."

After my son was born, I received an Evite for a party entitled "Please don't lick the baby." Further instructions asked guests to wash their hands before arrival and not to touch the baby anyway. This seemed sensible. Parenthood can make anyone a hormonal germophobe, and I was no different. I had visitors apply botanical hand sanitizer (we lived in San Francisco, where there was hippie Purell) and remove their shoes at the door. Yet despite my vigilance, my son grew into an allergic toddler. His eyes swelled shut, his bottom turned red, and his body erupted with hives after exposure to a litany of foods, dust, pollen, and even the housecat he was raised with. Doctors warned me to prepare for a lifetime of severe immune dysfunction.

The devastating irony is that the rise of diseases of inflammation in children—often called "modern plagues"—is most likely not caused by picking up the pathogens we fear. Rather, it's the result of not being exposed to the microbes that are key to maturing immunity. And how we enter the world determines our first colonizers.

In the birth canal, babies acquire *Lactobacillus,* which helps them digest milk and begins the process of lowering the gut's pH to the normal range. But babies born by cesarean miss out. Studies show they instead often end up with bacteria that are commonly found on the skin (sometimes not even the mother's), such as *Staphylococcus*—and in the case of one neonatal intensive care unit, antibiotic- and disinfectant-resistant bacteria. Abnormal colonization may explain why C-section babies seem to have a heightened risk for obesity, allergies, and asthma, which are linked to gut inflammation.

My son was not a C-section baby. But he did grow up in an apartment that might have been too clean. According to one theory, environmental exposures contribute to our development after birth, and recent studies seem to back that up. They suggest germs might actually help prevent children from developing various maladies.

"A house with a more bacteria-rich environment is a healthier one," says Susan Lynch, a microbiologist at the University of California at San Francisco. Her group profiled 104 infants inside their homes and found that the babies exposed to house dust with the greatest bacterial diversity before age one were the least likely to have asthma symptoms as three-year-olds. In addition to mouse and cockroach droppings, the dust was heavily colonized with microbes found in a healthy Western gut. Toddlers exposed to fewer types of bacteria, on the other hand, turned into hyperallergic wheezers. "We found that in homes with very little bacterial diversity," she says, "there was a very large number of fungi present."

Because studies show pet exposure might protect kids from allergies, Lynch also fed young mice meals from homes with germ-rich dogs. The mice grew up to be less allergic than those used as controls. She isolated one of their gut microbes, *Lactobacillus johnsonii,* and fed it to more mice. They were protected, too, but less so. Lynch suspects that *L. johnsonii* is a "keystone" species: it has an outsize role in determining which microbes move in and how they behave—guiding the immune response.

I've met Lynch before, when my son was morphing into one of her asthmatic superwheezers. She helped line up a medical referral. "How's he doing?" she now asks.

I tell her we moved to Oakland, where I countered my son's rather unscientific medical diagnosis of "allergic gut" with an equally unscientific prescription of dirt, dogs, chickens, and cultured foods. After school he tends to his bean tepee, and grows the strawberries he once couldn't eat. A fine sprinkling of soil often rings his mouth, like cookie crumbs. Surprisingly, most of his allergies have disappeared.

"He sounds like a perfect case study," Lynch says, completely nonplussed. "I would have liked to have gotten samples from him before and after. My guess is that his microbiome looks more like a normal gut." Lynch recently left San Francisco, too; it turns out we're neighbors. "We have a great picture of our 10-month-old daughter eating soil off a rock," she says.

In a remote corner of Northern California, on a steep slope of knotty oaks, sulfur and steam rise in plumes from Wilbur Hot Springs. It's the perfect place, says Eisen, to investigate the ghost limbs on the tree of life, the ones that contain multitudes of microbes we haven't yet identified. This microbial dark matter, as he calls it, is best pursued in isolated locales, such as deep mines and underground aquifers—or a nearby pool of absinthe-colored spring water, by which a sunbather lounges in a broad hat, and not much else.

This place is weird, and it is Eisen's milieu. He enters a creaky wooden shack, where water from a spigot feeds the pool. His colleagues from the Department of Energy's Joint Genome Institute, where he is an adjunct scientist, were here months earlier with collection jars. They were taking the waters, to echo an old phrase referring to the devotees of spa towns—only quite literally. They took samples back to the lab, where they amplified the microbial DNA a billionfold.

As we hike along a creek toward the source water, Eisen is in a good mood. The view's nice; the chaparral smells great. Here, he makes his final case for microbial diversity: dark matter is special, he tells me. By 2009 scientists had sequenced the DNA of only about a thousand microbes, those important to medicine or with clear applications. They mainly came from the same three branches of the evolutionary tree. And so Eisen led a team that set out to sequence a thousand more, with an emphasis on "neglected" species. The work has begun to fill in the tree with many more branches, revealing how microbes evolved and how species are related.

Ultimately, Eisen hopes, this knowledge will provide "a field guide to all microbes, including what is normally seen in the built environment." Much of the DNA found in recent studies lacks context. In addition, many microbes have genes with completely unknown functions. Finding similar genes on different branches could explain what they do—and eventually help us select microbes to create healthier surroundings. Emily Landon, an epidemiologist at the University of Chicago, envisions one day replacing antimicrobial paint with probiotics-infused walls. She calls it a fecal transplant for the built environment, wherein we infuse a space with beneficial bacteria that outcompete harmful ones. Or somewhere in Lynch's pile of anonymous DNA could be a clue to a microbe that eliminates my son's remaining allergy, to our cat.

Near the ruins of a bathhouse, milky bubbles well up from an aquifer. Garishly colored films have formed on rocks poking out of the water. "This is pretty awesome," Eisen says, wading toward a red-and-purple blob. "That's a nice mat. Touch that." As he inspects the photosynthetic bacteria, a cloud of tiny winged insects hovers at his ankles. These bugs, too, are taking the waters. Chances are they evolved to be at home with their own set of microbes. As we have.

OLIVER SACKS

My Periodic Table

FROM *The New York Times*

I LOOK FORWARD EAGERLY, almost greedily, to the weekly arrival of journals like *Nature* and *Science,* and turn at once to articles on the physical sciences—not, as perhaps I should, to articles on biology and medicine. It was the physical sciences that provided my first enchantment as a boy.

In a recent issue of *Nature,* there was a thrilling article by the Nobel Prize–winning physicist Frank Wilczek on a new way of calculating the slightly different masses of neutrons and protons. The new calculation confirms that neutrons are very slightly heavier than protons—the ratio of their masses being 939.56563 to 938.27231—a trivial difference, one might think, but if it were otherwise the universe as we know it could never have developed. The ability to calculate this, Dr. Wilczek wrote, "encourages us to predict a future in which nuclear physics reaches the level of precision and versatility that atomic physics has already achieved"—a revolution that, alas, I will never see.

Francis Crick was convinced that "the hard problem"—understanding how the brain gives rise to consciousness—would be solved by 2030. "You will see it," he often said to my neuroscientist friend Ralph, "and you may, too, Oliver, if you live to my age." Crick lived to his late 80s, working and thinking about consciousness till the last. Ralph died prematurely, at age 52, and now I am terminally ill, at the age of 82. I have to say that I am not too exercised by "the hard problem" of consciousness—indeed, I do not see it as a problem at all; but I am sad that I will not see the new nuclear physics that Dr. Wilczek envisages, nor a thousand other breakthroughs in the physical and biological sciences.

A few weeks ago, in the country, far from the lights of the city, I saw the entire sky "powdered with stars" (in Milton's words); such a sky, I imagined, could be seen only on high, dry plateaus like that of Atacama in Chile (where some of the world's most powerful telescopes are). It was this celestial splendor that suddenly made me realize how little time, how little life, I had left. My sense of the heavens' beauty, of eternity, was inseparably mixed for me with a sense of transience—and death.

I told my friends Kate and Allen, "I would like to see such a sky again when I am dying."

"We'll wheel you outside," they said.

I have been comforted, since I wrote in February about having metastatic cancer, by the hundreds of letters I have received, the expressions of love and appreciation, and the sense that (despite everything) I may have lived a good and useful life. I remain very glad and grateful for all this—yet none of it hits me as did that night sky full of stars.

I have tended since early boyhood to deal with loss—losing people dear to me—by turning to the nonhuman. When I was sent away to a boarding school as a child of 6, at the outset of World War II, numbers became my friends; when I returned to London at 10, the elements and the periodic table became my companions. Times of stress throughout my life have led me to turn, or return, to the physical sciences, a world where there is no life, but also no death.

And now, at this juncture, when death is no longer an abstract concept, but a presence—an all-too-close, not-to-be-denied presence—I am again surrounding myself, as I did when I was a boy, with metals and minerals, little emblems of eternity. At one end of my writing table, I have element 81 in a charming box, sent to me by element-friends in England: it says, HAPPY THALLIUM BIRTHDAY, a souvenir of my 81st birthday last July; then, a realm devoted to lead, element 82, for my just-celebrated 82nd birthday earlier this month. Here, too, is a little lead casket, containing element 90, thorium, crystalline thorium, as beautiful as diamonds, and, of course, radioactive—hence the lead casket.

At the start of the year, in the weeks after I learned that I had cancer, I *felt* pretty well, despite my liver being half-occupied by metastases. When the cancer in my liver was treated in February by the injection of tiny beads into the hepatic arteries—a procedure called embolization—I felt awful for a couple of weeks but then superwell, charged with physical and mental energy. (The metastases had almost all been wiped out by the embolization.) I had been given not a remission, but an intermission, a time to deepen friendships, to see patients, to write, and to travel back to my homeland, England. People could scarcely believe at this time that I had a terminal condition, and I could easily forget it myself.

This sense of health and energy started to decline as May moved into June, but I was able to celebrate my 82nd birthday in style. (Auden used to say that one should *always* celebrate one's birthday, no matter how one felt.) But now, I have some nausea and loss of appetite; chills in the day, sweats at night; and, above all, a pervasive tiredness, with sudden exhaustion if I overdo things. I continue to swim daily, but more slowly now, as I am beginning to feel a little short of breath. I could deny it before, but I *know* I am ill now. A CT scan on July 7 confirmed that the metastases had not only regrown in my liver but had now spread beyond it as well.

I started a new sort of treatment—immunotherapy—last week. It is not without its hazards, but I hope it will give me a few more good months. But before beginning this, I wanted to have a little fun: a trip to North Carolina to see the wonderful lemur research center at Duke University. Lemurs are close to the ancestral stock from which all primates arose, and I am happy to think that one of my own ancestors, 50 million years ago, was a little tree-dwelling creature not so dissimilar to the lemurs of today. I love their leaping vitality, their inquisitive nature.

Next to the circle of lead on my table is the land of bismuth: naturally occurring bismuth from Australia; little limousine-shaped ingots of bismuth from a mine in Bolivia; bismuth slowly cooled from a melt to form beautiful iridescent crystals terraced like a Hopi village; and, in a nod to Euclid and the beauty of geometry, a cylinder and a sphere made of bismuth.

Bismuth is element 83. I do not think I will see my 83rd birthday, but I feel there is something hopeful, something encouraging, about having "83" around. Moreover, I have a soft spot for bismuth, a modest gray metal, often unregarded, ignored, even by metal lovers. My feeling as a doctor for the mistreated or marginalized extends into the inorganic world and finds a parallel in my feeling for bismuth.

I almost certainly will not see my polonium (84th) birthday, nor would I want any polonium around, with its intense, murderous radioactivity. But then, at the other end of my table—my periodic table—I have a beautifully machined piece of beryllium (element 4) to remind me of my childhood, and of how long ago my soon-to-end life began.

KATHRYN SCHULZ

The Really Big One

FROM *The New Yorker*

WHEN THE 2011 EARTHQUAKE and tsunami struck Tohoku, Japan, Chris Goldfinger was 200 miles away, in the city of Kashiwa, at an international meeting on seismology. As the shaking started, everyone in the room began to laugh. Earthquakes are common in Japan—that one was the third of the week—and the participants were, after all, at a seismology conference. Then everyone in the room checked the time.

Seismologists know that how long an earthquake lasts is a decent proxy for its magnitude. The 1989 earthquake in Loma Prieta, California, which killed 63 people and caused $6 billion worth of damage, lasted about 15 seconds and had a magnitude of 6.9. A 30-second earthquake generally has a magnitude in the mid-7s. A minute-long quake is in the high 7s, a two-minute quake has entered the 8s, and a three-minute quake is in the high 8s. By four minutes, an earthquake has hit magnitude 9.0.

When Goldfinger looked at his watch, it was quarter to three. The conference was wrapping up for the day. He was thinking about sushi. The speaker at the lectern was wondering if he should carry on with his talk. The earthquake was not particularly strong. Then it ticked past the 60-second mark, making it longer than the others that week. The shaking intensified. The seats in the conference room were small plastic desks with wheels. Goldfinger, who is tall and solidly built, thought, *No way am I crouching under one of those for cover.* At a minute and a half, everyone in the room got up and went outside.

It was March. There was a chill in the air, and snow flurries, but no snow on the ground. Nor, from the feel of it, was there ground on the ground. The earth snapped and popped and rippled. It was, Goldfinger thought, like driving through rocky terrain in a vehicle with no shocks, if both the vehicle and the terrain were also on a raft in high seas. The quake passed the two-minute mark. The trees, still hung with the previous autumn's dead leaves, were making a strange rattling sound. The flagpole atop the building he and his colleagues had just vacated was whipping through an arc of 40 degrees. The building itself was base-isolated, a seismic-safety technology in which the body of a structure rests on movable bearings rather than directly on its foundation. Goldfinger lurched over to take a look. The base was lurching, too, back and forth a foot at a time, digging a trench in the yard. He thought better of it, and lurched away. His watch swept past the three-minute mark and kept going.

Oh, shit, Goldfinger thought, although not in dread, at first: in amazement. For decades, seismologists had believed that Japan could not experience an earthquake stronger than magnitude 8.4. In 2005, however, at a conference in Hokudan, a Japanese geologist named Yasutaka Ikeda had argued that the nation should expect a magnitude 9.0 in the near future—with catastrophic consequences, because Japan's famous earthquake-and-tsunami preparedness, including the height of its seawalls, was based on incorrect science. The presentation was met with polite applause and thereafter largely ignored. Now, Goldfinger realized as the shaking hit the four-minute mark, the planet was proving the Japanese Cassandra right.

For a moment, that was pretty cool: a real-time revolution in earthquake science. Almost immediately, though, it became extremely uncool, because Goldfinger and every other seismologist standing outside in Kashiwa knew what was coming. One of them pulled out a cell phone and started streaming videos from the Japanese broadcasting station NHK, shot by helicopters that had flown out to sea soon after the shaking started. Thirty minutes after Goldfinger first stepped outside, he watched the tsunami roll in, in real time, on a two-inch screen.

In the end, the magnitude-9.0 Tohoku earthquake and subsequent tsunami killed more than 18,000 people, devastated northeast Japan, triggered the meltdown at the Fukushima power plant, and cost an estimated $220 billion. The shaking earlier in the week turned out to be the foreshocks of the largest earthquake in the nation's recorded history. But for Chris Goldfinger, a paleoseismologist at Oregon State University and one of the world's leading experts on a little-known fault line, the main quake was itself a kind of foreshock: a preview of another earthquake still to come.

Most people in the United States know just one fault line by name: the San Andreas, which runs nearly the length of California and is perpetually rumored to be on the verge of unleashing "the big one." That rumor is misleading, no matter what the San Andreas ever does. Every fault line has an upper limit to its potency, determined by its length and width, and by how far it can slip. For the San Andreas, one of the most extensively studied and best-understood fault lines in the world, that upper limit is roughly an 8.2—a powerful earthquake, but, because the Richter scale is logarithmic, only 6 percent as strong as the 2011 event in Japan.

Just north of the San Andreas, however, lies another fault line. Known as the Cascadia subduction zone, it runs for 700 miles off the coast of the Pacific Northwest, beginning near Cape Mendocino, California, continuing along Oregon and Washington, and terminating around Vancouver Island, Canada. The "Cascadia" part of its name comes from the Cascade Range, a chain of volcanic mountains that follow the same course a hundred or so miles inland. The "subduction zone" part refers to a region of the planet where one tectonic plate is sliding underneath (subducting) another. Tectonic plates are those slabs of mantle and crust that, in their epochs-long drift, rearrange the earth's continents and oceans. Most of the time, their movement is slow, harmless, and all but undetectable. Occasionally, at the borders where they meet, it is not.

Take your hands and hold them palms down, middle fingertips touching. Your right hand represents the North American tectonic plate, which bears on its back, among other things, our entire continent, from One World Trade Center to the Space Needle, in Seattle. Your left hand represents an oceanic plate called Juan de Fuca, 90,000 square miles in size. The place where they meet is the Cascadia subduction zone. Now slide your left hand under your right one. That is what the Juan de Fuca plate is doing: slipping steadily beneath North America. When you try it, your right hand will slide up your left arm, as if you were pushing up your sleeve. That is what North America is not doing. It is stuck, wedged tight against the surface of the other plate.

Without moving your hands, curl your right knuckles up, so that they point toward the ceiling. Under pressure from Juan de Fuca, the stuck edge of North America is bulging upward and compressing eastward, at the rate of, respectively, 3 to 4 millimeters and 30 to 40 millimeters a year. It can do so for quite some time, because, as continent stuff goes, it is young, made of rock that is still relatively elastic. (Rocks, like us, get stiffer as they age.) But it cannot do so indefinitely. There is a backstop—the craton, that ancient unbudgeable mass at the center of the continent—and, sooner or later, North America will rebound like a spring. If, on that occasion, only the southern part of the Cascadia subduction zone gives way—your first two fingers, say—the magnitude of the resulting quake will be somewhere between 8.0 and 8.6. *That's* the big one. If the entire zone gives way at once, an event that seismologists call a full-margin rupture, the magnitude will be somewhere between 8.7 and 9.2. That's the very big one.

Flick your right fingers outward, forcefully, so that your hand flattens back down again. When the next very big earthquake hits, the northwest edge of the continent, from California to Canada and the continental shelf to the Cascades, will drop by as much as 6 feet and rebound 30 to 100 feet to the west—losing, within minutes, all the elevation and compression it has gained over centuries. Some of that shift will take place beneath the ocean, displacing a colossal quantity of seawater. (Watch what your fingertips do when you flatten your hand.) The water will surge upward into a huge hill, then promptly collapse. One side will rush west, toward Japan. The other side will rush east, in a 700-mile liquid wall that will reach the Northwest coast, on average, 15 minutes after the earthquake begins. By the time the shaking has ceased and the tsunami has receded, the region will be unrecognizable. Kenneth Murphy, who directs the Federal Emergency Management Agency's Region X, the division responsible for Oregon, Washington, Idaho, and Alaska, says, "Our operating assumption is that everything west of Interstate 5 will be toast."

In the Pacific Northwest, everything west of Interstate 5 covers some 140,000 square miles, including Seattle, Tacoma, Portland, Eugene, Salem (the capital city of Oregon), Olympia (the capital of Washington), and some seven million people. When the next full-margin rupture happens, that region will suffer the worst natural disaster in the history of North America. Roughly 3,000 people died in San Francisco's 1906 earthquake. Almost 2,000 died in Hurricane Katrina. Almost 300 died in Hurricane Sandy. FEMA projects that nearly 13,000 people will die in the Cascadia earthquake and tsunami. Another 27,000 will be injured, and the agency expects that it will need to provide shelter for 1 million displaced people, and food and water for another 2.5 million. "This is one time that I'm hoping all the science is wrong, and it won't happen for another thousand years," Murphy says.

In fact, the science is robust, and one of the chief scientists behind it is Chris Goldfinger. Thanks to work done by him and his colleagues, we now know that the odds of the big Cascadia earthquake happening in the next 50 years are roughly 1 in 3. The odds of the very big one are roughly 1 in 10. Even those numbers do not fully reflect the danger—or, more to the point, how unprepared the Pacific Northwest is to face it. The truly worrisome figures in this story are these: 30 years ago, no one knew that the Cascadia subduction zone had ever produced a major earthquake; 45 years ago, no one even knew it existed.

In May of 1804 Meriwether Lewis and William Clark, together with their Corps of Discovery, set off from St. Louis on America's first official cross-country expedition. Eighteen months later, they reached the Pacific Ocean and made camp near the present-day town of Astoria, Oregon. The United States was, at the time, 29 years old. Canada was not yet a country. The continent's far expanses were so unknown to its white explorers that Thomas Jefferson, who commissioned the journey, thought that the men would come across woolly mammoths. Native Americans had lived in the Northwest for millennia, but they had no written language, and the many things to which the arriving Europeans subjected them did not include seismological inquiries. The newcomers took the land they encountered at face value, and at face value it was a find: vast, cheap, temperate, fertile, and, to all appearances, remarkably benign.

A century and a half elapsed before anyone had any inkling that the Pacific Northwest was not a quiet place but a place in a long period of quiet. It took another 50 years to uncover and interpret the region's seismic history. Geology, as even geologists will tell you, is not normally the sexiest of disciplines; it hunkers down with earthly stuff while the glory accrues to the human and the cosmic —to genetics, neuroscience, physics. But, sooner or later, every field has its field day, and the discovery of the Cascadia subduction zone stands as one of the greatest scientific detective stories of our time.

The first clue came from geography. Almost all of the world's most powerful earthquakes occur in the Ring of Fire, the volcanically and seismically volatile swath of the Pacific that runs from New Zealand up through Indonesia and Japan, across the ocean to Alaska, and down the west coast of the Americas to Chile. Japan, 2011, magnitude 9.0; Indonesia, 2004, magnitude 9.1; Alaska, 1964, magnitude 9.2; Chile, 1960, magnitude 9.5—not until the late 1960s, with the rise of the theory of plate tectonics, could geologists explain this pattern. The Ring of Fire, it turns out, is really a ring of subduction zones. Nearly all the earthquakes in the region are caused by continental plates getting stuck on oceanic plates —as North America is stuck on Juan de Fuca—and then getting abruptly unstuck. And nearly all the volcanoes are caused by the oceanic plates sliding deep beneath the continental ones, eventually reaching temperatures and pressures so extreme that they melt the rock above them.

The Pacific Northwest sits squarely within the Ring of Fire. Off its coast, an oceanic plate is slipping beneath a continental one. Inland, the Cascade volcanoes mark the line where, far below, the Juan de Fuca plate is heating up and melting everything above it. In other words, the Cascadia subduction zone has, as Goldfinger put it, "all the right anatomical parts." Yet not once in recorded history has it caused a major earthquake—or, for that matter, any quake to speak of. By contrast, other subduction zones produce major earthquakes occasionally and minor ones all the time: magnitude 5.0, magnitude 4.0, magnitude why are the neighbors moving their sofa at midnight. You can scarcely spend a week in Japan without feeling this sort of earthquake. You can spend a lifetime in many parts of the Northwest—several, in fact, if you had them to spend—and not feel so much as a quiver. The question facing geologists in the 1970s was whether the Cascadia subduction zone had ever broken its eerie silence.

In the late 1980s, Brian Atwater, a geologist with the United

States Geological Survey, and a graduate student named David Yamaguchi found the answer, and another major clue in the Cascadia puzzle. Their discovery is best illustrated in a place called the ghost forest, a grove of western red cedars on the banks of the Copalis River, near the Washington coast. When I paddled out to it last summer, with Atwater and Yamaguchi, it was easy to see how it got its name. The cedars are spread out across a low salt marsh on a wide northern bend in the river, long dead but still standing. Leafless, branchless, barkless, they are reduced to their trunks and worn to a smooth silver-gray, as if they had always carried their own tombstones inside them.

What killed the trees in the ghost forest was salt water. It had long been assumed that they died slowly, as the sea level around them gradually rose and submerged their roots. But, by 1987, Atwater, who had found in soil layers evidence of sudden land subsidence along the Washington coast, suspected that that was backward—that the trees had died quickly when the ground beneath them plummeted. To find out, he teamed up with Yamaguchi, a specialist in dendrochronology, the study of growth-ring patterns in trees. Yamaguchi took samples of the cedars and found that they had died simultaneously: in tree after tree, the final rings dated to the summer of 1699. Since trees do not grow in the winter, he and Atwater concluded that sometime between August of 1699 and May of 1700 an earthquake had caused the land to drop and killed the cedars. That time frame predated by more than 100 years the written history of the Pacific Northwest—and so, by rights, the detective story should have ended there.

But it did not. If you travel 5,000 miles due west from the ghost forest, you reach the northeast coast of Japan. As the events of 2011 made clear, that coast is vulnerable to tsunamis, and the Japanese have kept track of them since at least AD 599. In that 1,400-year history, one incident has long stood out for its strangeness. On the 8th day of the 12th month of the 12th year of the Genroku era, a 600-mile-long wave struck the coast, leveling homes, breaching a castle moat, and causing an accident at sea. The Japanese understood that tsunamis were the result of earthquakes, yet no one felt the ground shake before the Genroku event. The wave had no discernible origin. When scientists began studying it, they called it an orphan tsunami.

Finally, in a 1996 article in *Nature*, a seismologist named Kenji

Satake and three colleagues, drawing on the work of Atwater and Yamaguchi, matched that orphan to its parent—and thereby filled in the blanks in the Cascadia story with uncanny specificity. At approximately nine o'clock at night on January 26, 1700, a magnitude-9.0 earthquake struck the Pacific Northwest, causing sudden land subsidence, drowning coastal forests, and, out in the ocean, lifting up a wave half the length of a continent. It took roughly 15 minutes for the eastern half of that wave to strike the Northwest coast. It took 10 hours for the other half to cross the ocean. It reached Japan on January 27, 1700: by the local calendar, the 8th day of the 12th month of the 12th year of Genroku.

Once scientists had reconstructed the 1700 earthquake, certain previously overlooked accounts also came to seem like clues. In 1964 Chief Louis Nookmis, of the Huu-ay-aht First Nation, in British Columbia, told a story, passed down through seven generations, about the eradication of Vancouver Island's Pachena Bay people. "I think it was at nighttime that the land shook," Nookmis recalled. According to another tribal history, "They sank at once, were all drowned; not one survived." A hundred years earlier, Billy Balch, a leader of the Makah tribe, recounted a similar story. Before his own time, he said, all the water had receded from Washington State's Neah Bay, then suddenly poured back in, inundating the entire region. Those who survived later found canoes hanging from the trees. In a 2005 study, Ruth Ludwin, then a seismologist at the University of Washington, together with nine colleagues, collected and analyzed Native American reports of earthquakes and saltwater floods. Some of those reports contained enough information to estimate a date range for the events they described. On average, the midpoint of that range was 1701.

It does not speak well of European Americans that such stories counted as evidence for a proposition only after that proposition had been proved. Still, the reconstruction of the Cascadia earthquake of 1700 is one of those rare natural puzzles whose pieces fit together as tectonic plates do not: perfectly. It is wonderful science. It was wonderful *for* science. And it was terrible news for the millions of inhabitants of the Pacific Northwest. As Goldfinger put it, "In the late eighties and early nineties, the paradigm shifted to 'uh-oh.'"

Goldfinger told me this in his lab at Oregon State, a low prefab building that a passing English major might reasonably mistake for the maintenance department. Inside the lab is a walk-in freezer. Inside the freezer are floor-to-ceiling racks filled with cryptically labeled tubes, four inches in diameter and five feet long. Each tube contains a core sample of the seafloor. Each sample contains the history, written in seafloor-ese, of the past 10,000 years. During subduction-zone earthquakes, torrents of land rush off the continental slope, leaving a permanent deposit on the bottom of the ocean. By counting the number and the size of deposits in each sample, then comparing their extent and consistency along the length of the Cascadia subduction zone, Goldfinger and his colleagues were able to determine how much of the zone has ruptured, how often, and how drastically.

Thanks to that work, we now know that the Pacific Northwest has experienced 41 subduction-zone earthquakes in the past 10,000 years. If you divide 10,000 by 41, you get 243, which is Cascadia's recurrence interval: the average amount of time that elapses between earthquakes. That timespan is dangerous both because it is too long—long enough for us to unwittingly build an entire civilization on top of our continent's worst fault line—and because it is not long enough. Counting from the earthquake of 1700, we are now 315 years into a 243-year cycle.

It is possible to quibble with that number. Recurrence intervals are averages, and averages are tricky: 10 is the average of 9 and 11, but also of 18 and 2. It is not possible, however, to dispute the scale of the problem. The devastation in Japan in 2011 was the result of a discrepancy between what the best science predicted and what the region was prepared to withstand. The same will hold true in the Pacific Northwest—but here the discrepancy is enormous. "The science part is fun," Goldfinger says. "And I love doing it. But the gap between what we know and what we should do about it is getting bigger and bigger, and the action really needs to turn to responding. Otherwise, we're going to be hammered. I've been through one of these massive earthquakes in the most seismically prepared nation on earth. If that was Portland"—Goldfinger finished the sentence with a shake of his head before he finished it with words. "Let's just say I would rather not be here."

The first sign that the Cascadia earthquake has begun will be a compressional wave, radiating outward from the fault line. Compressional waves are fast-moving, high-frequency waves, audible to dogs and certain other animals but experienced by humans only as a sudden jolt. They are not very harmful, but they are potentially very useful, since they travel fast enough to be detected by sensors 30 to 90 seconds ahead of other seismic waves. That is enough time for earthquake early-warning systems, such as those in use throughout Japan, to automatically perform a variety of lifesaving functions: shutting down railways and power plants, opening elevators and firehouse doors, alerting hospitals to halt surgeries, and triggering alarms so that the general public can take cover. The Pacific Northwest has no early-warning system. When the Cascadia earthquake begins, there will be, instead, a cacophony of barking dogs and a long, suspended, what-was-that moment before the surface waves arrive. Surface waves are slower, lower-frequency waves that move the ground both up and down and side to side: the shaking, starting in earnest.

Soon after that shaking begins, the electrical grid will fail, likely everywhere west of the Cascades and possibly well beyond. If it happens at night, the ensuing catastrophe will unfold in darkness. In theory, those who are at home when it hits should be safest; it is easy and relatively inexpensive to seismically safeguard a private dwelling. But, lulled into nonchalance by their seemingly benign environment, most people in the Pacific Northwest have not done so. That nonchalance will shatter instantly. So will everything made of glass. Anything indoors and unsecured will lurch across the floor or come crashing down: bookshelves, lamps, computers, canisters of flour in the pantry. Refrigerators will walk out of kitchens, unplugging themselves and toppling over. Water heaters will fall and smash interior gas lines. Houses that are not bolted to their foundations will slide off—or, rather, they will stay put, obeying inertia, while the foundations, together with the rest of the Northwest, jolt westward. Unmoored on the undulating ground, the homes will begin to collapse.

Across the region, other, larger structures will also start to fail. Until 1974, the state of Oregon had no seismic code, and few places in the Pacific Northwest had one appropriate to a magnitude-9.0 earthquake until 1994. The vast majority of buildings in the region were constructed before then. Ian Madin, who directs the Oregon Department of Geology and Mineral Industries (DOGAMI), estimates that 75 percent of all structures in the state are not designed to withstand a major Cascadia quake. FEMA calculates that, across the region, something on the order of a million buildings—more than 3,000 of them schools—will collapse or be compromised in the earthquake. So will half of all highway bridges, 15 of the 17 bridges spanning Portland's two rivers, and two-thirds of railways and airports; also, one-third of all fire stations, half of all police stations, and two-thirds of all hospitals.

Certain disasters stem from many small problems conspiring to cause one very large problem. For want of a nail, the war was lost; for 15 independently insignificant errors, the jetliner was lost. Subduction-zone earthquakes operate on the opposite principle: one enormous problem causes many other enormous problems. The shaking from the Cascadia quake will set off landslides throughout the region—up to 30,000 of them in Seattle alone, the city's emergency-management office estimates. It will also induce a process called liquefaction, whereby seemingly solid ground starts behaving like a liquid, to the detriment of anything on top of it. Fifteen percent of Seattle is built on liquefiable land, including 17 daycare centers and the homes of some 34,500 people. So is Oregon's critical energy-infrastructure hub, a six-mile stretch of Portland through which flows 90 percent of the state's liquid fuel and which houses everything from electrical substations to natural-gas terminals. Together, the sloshing, sliding, and shaking will trigger fires, flooding, pipe failures, dam breaches, and hazardous-material spills. Any one of these second-order disasters could swamp the original earthquake in terms of cost, damage, or casualties—and one of them definitely will. Four to six minutes after the dogs start barking, the shaking will subside. For another few minutes, the region, upended, will continue to fall apart on its own. Then the wave will arrive, and the real destruction will begin.

Among natural disasters, tsunamis may be the closest to being completely unsurvivable. The only likely way to outlive one is not to be there when it happens: to steer clear of the vulnerable area in the first place, or get yourself to high ground as fast as possible. For the 71,000 people who live in Cascadia's inundation zone, that will mean evacuating in the narrow window after one disaster ends and before another begins. They will be notified to do so only by the earthquake itself—"a vibrate-alert system," Kevin Cupples, the city planner for the town of Seaside, Oregon, jokes—and they are urged to leave on foot, since the earthquake will render roads impassable. Depending on location, they will have between 10 and 30 minutes to get out. That timeline does not allow for finding a flashlight, tending to an earthquake injury, hesitating amid the ruins of a home, searching for loved ones, or being a Good Samaritan. "When that tsunami is coming, you run," Jay Wilson, the chair of the Oregon Seismic Safety Policy Advisory Commission (OSSPAC), says. "You protect yourself, you don't turn around, you don't go back to save anybody. You run for your life."

The time to save people from a tsunami is before it happens, but the region has not yet taken serious steps toward doing so. Hotels and businesses are not required to post evacuation routes or to provide employees with evacuation training. In Oregon, it has been illegal since 1995 to build hospitals, schools, firehouses, and police stations in the inundation zone, but those which are already in it can stay, and any other new construction is permissible: energy facilities, hotels, retirement homes. In those cases, builders are required only to consult with DOGAMI about evacuation plans. "So you come in and sit down," Ian Madin says. "And I say, 'That's a stupid idea.' And you say, 'Thanks. Now we've consulted.'"

These lax safety policies guarantee that many people inside the inundation zone will not get out. Twenty-two percent of Oregon's coastal population is 65 or older. Twenty-nine percent of the state's population is disabled, and that figure rises in many coastal counties. "We can't save them," Kevin Cupples says. "I'm not going to sugarcoat it and say, 'Oh, yeah, we'll go around and check on the elderly.' No. We won't." Nor will anyone save the tourists. Washington State Park properties within the inundation zone see an average of 17,029 guests a day. Madin estimates that up to 150,000 people visit Oregon's beaches on summer weekends. "Most of them won't have a clue as to how to evacuate," he says. "And the beaches are the hardest place to evacuate from."

Those who cannot get out of the inundation zone under their own power will quickly be overtaken by a greater one. A grown man is knocked over by ankle-deep water moving at 6.7 miles an hour. The tsunami will be moving more than twice that fast when it arrives. Its height will vary with the contours of the coast, from 20 feet to more than 100 feet. It will not look like a Hokusai-style wave, rising up from the surface of the sea and breaking from above. It will look like the whole ocean, elevated, overtaking land. Nor will it be made only of water—not once it reaches the shore. It will be a five-story deluge of pickup trucks and doorframes and cinder blocks and fishing boats and utility poles and everything else that once constituted the coastal towns of the Pacific Northwest.

To see the full scale of the devastation when that tsunami recedes, you would need to be in the International Space Station. The inundation zone will be scoured of structures from California to Canada. The earthquake will have wrought its worst havoc west of the Cascades but caused damage as far away as Sacramento, California—as distant from the worst-hit areas as Fort Wayne, Indiana, is from New York. FEMA expects to coordinate search-and-rescue operations across 100,000 square miles and in the waters off 453 miles of coastline. As for casualties: the figures I cited earlier —27,000 injured, almost 13,000 dead—are based on the agency's official planning scenario, which has the earthquake striking at 9:41 a.m. on February 6. If, instead, it strikes in the summer, when the beaches are full, those numbers could be off by a horrifying margin.

Wineglasses, antique vases, Humpty Dumpty, hip bones, hearts: what breaks quickly generally mends slowly, if at all. OSSPAC estimates that in the I-5 corridor it will take between 1 and 3 months after the earthquake to restore electricity, a month to a year to restore drinking water and sewer service, 6 months to a year to restore major highways, and 18 months to restore health-care facilities. On the coast, those numbers go up. Whoever chooses or has no choice but to stay there will spend 3 to 6 months without electricity, one to three years without drinking water and sewage systems, and three or more years without hospitals. Those estimates do not apply to the tsunami-inundation zone, which will remain all but uninhabitable for years.

How much all this will cost is anyone's guess; FEMA puts every number on its relief-and-recovery plan except a price. But whatever the ultimate figure—and even though U.S. taxpayers will cover 75 to 100 percent of the damage, as happens in declared disasters —the economy of the Pacific Northwest will collapse. Crippled by a lack of basic services, businesses will fail or move away. Many residents will flee as well. OSSPAC predicts a mass-displacement event and a long-term population downturn. Chris Goldfinger didn't want to be there when it happened. But by many metrics, it will be as bad or worse to be there afterward.

On the face of it, earthquakes seem to present us with problems of space: the way we live along fault lines, in brick buildings, in homes made valuable by their proximity to the sea. But, covertly, they also present us with problems of time. The earth is 4.5 billion years old, but we are a young species, relatively speaking, with an average individual allotment of 3 score years and 10. The brevity of our lives breeds a kind of temporal parochialism—an ignorance of or an indifference to those planetary gears which turn more slowly than our own.

This problem is bidirectional. The Cascadia subduction zone remained hidden from us for so long because we could not see deep enough into the past. It poses a danger to us today because we have not thought deeply enough about the future. That is no longer a problem of information; we now understand very well what the Cascadia fault line will someday do. Nor is it a problem of imagination. If you are so inclined, you can watch an earthquake destroy much of the West Coast this summer in Brad Peyton's *San Andreas,* while, in neighboring theaters, the world threatens to succumb to Armageddon by other means: viruses, robots, resource scarcity, zombies, aliens, plague. As those movies attest, we excel at imagining future scenarios, including awful ones. But such apocalyptic visions are a form of escapism, not a moral summons, and still less a plan of action. Where we stumble is in conjuring up grim futures in a way that helps to avert them.

That problem is not specific to earthquakes, of course. The Cascadia situation, a calamity in its own right, is also a parable for this age of ecological reckoning, and the questions it raises are ones that we all now face. How should a society respond to a looming crisis of uncertain timing but of catastrophic proportions? How can it begin to right itself when its entire infrastructure and culture developed in a way that leaves it profoundly vulnerable to natural disaster?

The last person I met with in the Pacific Northwest was Doug Dougherty, the superintendent of schools for Seaside, which lies almost entirely within the tsunami-inundation zone. Of the four schools that Dougherty oversees, with a total student population of 1,600, one is relatively safe. The others sit 5 to 15 feet above sea level. When the tsunami comes, they will be as much as 45 feet below it.

In 2009, Dougherty told me, he found some land for sale outside the inundation zone, and proposed building a new K–12 campus there. Four years later, to foot the $128 million bill, the district put up a bond measure. The tax increase for residents amounted to $2.16 per $1,000 of property value. The measure failed by 62 percent. Dougherty tried seeking help from Oregon's congressional delegation but came up empty. The state makes money available for seismic upgrades, but buildings within the inundation zone cannot apply. At present, all Dougherty can do is make sure that his students know how to evacuate.

Some of them, however, will not be able to do so. At an elementary school in the community of Gearhart, the children will be trapped. "They can't make it out from that school," Dougherty said. "They have no place to go." On one side lies the ocean; on the other, a wide, roadless bog. When the tsunami comes, the only place to go in Gearhart is a small ridge just behind the school. At its tallest, it is 45 feet high—lower than the expected wave in a full-margin earthquake. For now, the route to the ridge is marked by signs that say TEMPORARY TSUNAMI ASSEMBLY AREA. I asked Dougherty about the state's long-range plan. "There is no long-range plan," he said.

Dougherty's office is deep inside the inundation zone, a few blocks from the beach. All day long, just out of sight, the ocean rises up and collapses, spilling foamy overlapping ovals onto the shore. Eighty miles farther out, 10,000 feet below the surface of the sea, the hand of a geological clock is somewhere in its slow sweep. All across the region, seismologists are looking at their watches, wondering how long we have, and what we will do, before geological time catches up to our own.

GAURAV RAJ TELHAN

Begin Cutting

FROM *Virginia Quarterly Review*

> There exists an allegiance between the dead and the unborn of which we the living are merely the ligature.
> —Robert Pogue Harrison, *The Dominion of the Dead*

MONSTERS. The label was affixed to a display case that housed a collection of glass jars. Sealed with a rusted screw-top lid or damp wooden cork, they contained malformed fetuses preserved in fluid. The term, seemingly a cruel joke, was part of an outmoded medical lexicon and described a fetus with gross congenital anomalies. Shelves of these specimens lined the passageway at the south entrance of the University of Virginia School of Medicine's anatomy laboratory, displayed safely behind glass that spanned the height of the walls. Like many students before me, I passed by this grisly shrine during the first week of medical school and saw my own uneasy reflection in the glare of light against glass.

Taped to the double doors that led into the laboratory was a list of cadaver assignments by table number and cause of death: respiratory arrest, myocardial infarction, aortic aneurysm, pulmonary embolism, cerebrovascular accident. My cadaver lay at table number 10. Cause of death? Multi-infarct dementia. Death, that is to say, by madness.

Past the doors, the white cinder-block walls of the lab were lined with posters and diagrams portraying the intricate network of nerves, muscles, and blood vessels in the human body. In one corner, the chalk-white bones of a human skeleton hung from a metal frame. In another corner was a large biohazard bin that could hold several hundred pounds of dissected cadaver waste—about five bodies' worth. Everywhere in the bleached walls of the laboratory—the sterile linoleum flooring, the burnished metal of dissection tables, the zippered white bags used to veil the dead, the gleaming instruments used to cut them open—I saw the landscape of a story into which I was being written.

My lab partners and I put on our full-body aprons and white latex gloves, then approached the table that held our cadaver, cocooned still in its polyurethane casing. I leaned and gripped the table's lip to steady my trembling hands. Touching the table seemed to suggest that I was willing to unzip the body bag. I looked around and saw that other groups had already unwrapped their cadavers and begun the necessary shaving and cleaning of the body. Everyone at table 10 was anxious to begin, so I fumbled for the zipper at the rostral end of the bag, where the head lies, and pulled.

The bag opened, revealing a woman's body: anonymous, feeble, and atrophied, in its eighth or ninth decade. Staring at the table, I felt an uncomfortable awareness of my own body's temporality, and to distract myself I thought of the 16th-century French poets who practiced the art of *blason,* intimately cataloging the parts of a woman's body in verse. I tried to put myself in their position, intent on keeping a mental inventory of the cadaver's pieces.

First her name: Stella, Italian for "star." My lab partner offered the appellation, elevating the frail body skyward. Her hair was next, wispy and uneven, absent in patches, her scalp translucent with blue-green vessels snaking just beneath skin. Several moles and polyps peppered her pallid brow. Her skin was desiccated; her face bore a stony countenance, like that of a mountain gradually eroding. Her cheekbones jutted out at angles. Her lips were deeply fissured, enough to reveal the yellow teeth behind them. She smelled of formaldehyde. Her eyelids, not entirely shut, revealed the green of her irises, afloat in a white scleral sea.

It didn't take long to realize that this distraction wasn't going to work.

Twelve years have passed since that first day in the anatomy lab, and the image of that anonymous woman beneath the polyurethane veil still grips me. It lingered with me long after her body had been dissected and discarded. To unburden myself, I tried to get to know her better. Was she a willing donor? A stranger with no final resting ground? Was she a practical woman? An idealist? Did she die alone?

To begin piecing my answers together, I borrowed a page from negative theology—the logic used by religious scholars to describe God in terms of what she is not. It was my own *via negativa*.

The first step was to contact the Virginia State Anatomical Program (VSAP), an office established in 1919 as the state's sole agency authorized to receive donations of human bodies for scientific study. The State Anatomical Gift Act governs the logistics of anatomical donation to VSAP in the commonwealth. One of its mandates is a declaration of intent, an "Instrument of Anatomical Gift." The single-sided form requires less information than a job application: name, phone number, address, Social Security number, race, place of birth, parents' names, status of military service, and highest level of education. No references needed.

Maybe, like the form, Stella could be seen as an "instrument" —her right of personhood legally revoked and her body made a vehicle for what the declaration of intent calls "scientific study, teaching, research," and, more vaguely, "other purposes." She was a tool, then, meant to unearth knowledge. A tool that the state reserves the right to discard if deemed unacceptable, like a dull knife or crooked plane.

To this end, VSAP states that only intact bodies are eligible for scientific research. Because Stella was not physically damaged or disfigured, she was considered suitable for dissection. Medical students, after all, must learn the whole of normal anatomy; a body with a shattered skeleton or missing pancreas would be like a book with torn or missing pages. Fortunately for Stella, she was not mutilated in an accident and had no violent trauma—such as dagger wounds, mechanical dismemberment, damage from high-velocity projectiles, bombs, et cetera—that would have excluded her. Schools prefer symmetry.

Other things that Stella was not: Stella was not autopsied or embalmed. Stella was not in the beginning stages of decomposition. Stella was not suspected of having contagious and communicable diseases such as AIDS, hepatitis, or active tuberculosis. Stella did not have an antibiotic-resistant infection. She was neither exorbitantly obese nor emaciated. It was because of everything she was not, because of her ways of not-being, that Stella was suited ultimately to become our teacher.

Pointing a gloved finger at the area between Stella's legs, a classmate asked, "Who's gonna shave the hooch?" His crass humor was strangely alleviating. Distancing ourselves from the mass of flesh on the table (some might say dehumanizing it) made what came next easier. The body had to be shaved for visibility and ease of dissection, then washed to remove any extraneous debris. My lab partners and I turned to the only woman in the group, but she declined and pressed the plastic razor into my hand. Caught between its blades were the hairs of neighboring cadavers, already shaved.

I began at Stella's head, where the hair clung flimsily by the root and fell away at scarcely a touch of the blade. The blade's cutting edge, already dulled considerably, created nicks along the surface of her scalp. The deeper cuts offered glimpses of underlying bone.

I shaved the patch of hair between her breasts, rinsing the blade before moving carefully toward the mons pubis, labia majora, and perineum—translated literally, from the Latin, as the "mountain of the groin," the "great lips," and, from the Greek, as the "excreting part." When I finished, my lab mates scrubbed the body with wet sponges: first the ventral surface, then (with the whole group turning Stella onto her stomach) the dorsal surface. Specially designed tracks ran along either side of the dissection table, draining the watery gray waste into a plastic bucket that hung at Stella's feet.

Stella was only one of approximately 40 cadavers in the lab. With an average estimated weight of 180 pounds per cadaver, a total of 7,200 pounds of flesh flowed through the anatomy lab each year. In light of such staggering numbers, no single body seems so radical an integer.

And yet it was almost impossible to see these cadavers (contained though they were within the walls of the lab) as mere assignments, rather than sudden existential provocations. Comfort can be taken, I suppose, in knowing that medical students no longer have to find the bodies they dissect. In centuries past, they were forced not only to study the cadaver as a scientific object but also, in the act of removing it from the place where its death was quietly concealed, confront it as a lurid phenomenon.

Consider, for example, the study of human anatomy in 18th- and 19th-century Britain. Before passage of the Anatomy Act of 1832, which expanded the legal supply of cadavers to include all bodies unclaimed after death (not just the bodies of those who died in prison), medical students desperate for a body to study would take it upon themselves to exhume the dead. Such practices are the focus of a disputed 1896 article by Thomas Wakley, in the British medical journal the *Lancet,* that is detailed enough to be read as a do-it-yourself instruction book for raiding graves. In Wakley's account, a grave robber, or "resurrectionist," would "remove a square of turf, about eighteen or twenty inches in diameter" 15 or 20 feet away from the grave. He would then "commence to mine . . . [a] rough slanting tunnel some five yards long . . . to be constructed so as to impinge exactly on the coffin head." The grave robber would then extract the entire coffin. If he was lucky, "the end of the coffin was wrenched off with hooks while still in the shelter of the tunnel, and the scalp or feet of the corpse secured through the open end, and the body pulled out, leaving the coffin almost intact and unmoved."

Stella's hypothetical tunnel, to accommodate the girth of her abdomen and hips, would have to have been at least three feet in diameter. I also figure an additional foot to leave space for any maneuvering that might be required. And at least five yards of rope to drag her out.

Standing above her shaven body, I was surprised to be facing this moment with Stella at all. Just minutes before, I was packing my bag to leave the auditorium across the hall when the professor's instructions—"Begin cutting"—stopped me cold. I had thought dissection wouldn't start for another week. I wasn't ready. Before this, "dissection" had been mostly metaphorical, a lit-crit term. Now the implications of the word were almost paralyzing. I knew I could dissect a poem, but a body was another story.

I wasn't the only one with concerns. I could sense the minor panic of some classmates, but it was muted by the zeal of a larger contingent rushing to get at it. Their enthusiasm was welcomed, expected even. "Anatomy is power," a professor said briskly as we entered the lab. It was clear: the timid were at a disadvantage. We had better get ready, and quickly.

I had assumed this irrevocable moment would unfold differently, that there would be someone to help us understand what was at stake, what we might stand to gain or lose with that first cut. I had spent the year before medical school immersed in the work of physician and writer Robert Coles. In class I had recalled his insistence on the moral education of medical students. "All the time we . . . send moral signals to our students," Coles said, "we let them know (by how we teach what we teach) the kind of people we expect them to be." I once envisioned medical school being filled with professors and students like Coles, a Platonic academy where thinking and doing came together in a moral synthesis. But as I sat there in the auditorium, thinking about Coles and his belief in the power of our vocabularies to shape our moral imaginations, I wondered: *What are we today? Butchers?*

There was hardly any time to indulge these questions in the anatomy lab. Before I knew it, the razor in my hand was swapped for a scalpel. But the prospect of actually cutting into Stella threatened to force me into reckoning with my own vulnerability. As a countermeasure, I tried to make meaning out of matter.

My words came out shaky. "This cadaver," I said, "is a blueprint . . . for the future." I paused to look at my classmates, unsure if they wanted to hear any of this. I had already been slow shaving Stella's hair; this was only dragging things out more. "It's a map of . . . more than just these organs," I stammered, "but a guide to the bodies of the patients . . . the living ones we'll see one day." For years I had immersed myself in medical literature, reading physician-writers—Richard Selzer, Lewis Thomas, Abraham Verghese, William Carlos Williams—with an awe that affirmed my desire to live a life in medicine. But I struggled to reconcile their dazzling language with the reality on the table. And as my stilted monologue unfolded, one of several instructors pacing the lab quietly materialized from around a corner to deliver a perplexed, impatient look in my direction.

Still, it felt good to say something and offer at least some clumsy resistance to the subtle pressure to be silent, to name only the things that the lab manual instructed us to name, to see only what the lab manual told us to see. I suppose that what I had wanted to say was that if done right, the relationship we were cultivating with this body would shape our relationships with future patients. If we could stand in reverence before this mass of decomposing flesh, how could we not care deeply for the living, breathing bodies that would one day come under our care?

Washed clean and lying on her stomach, with her nose pressed against the table's steel surface, Stella was now in the proper "prone" position described in the lab manual. The dissection was organized by tissue layers. Our first task was to peel back the skin and subcutaneous tissue from the underlying muscle. With the help of a partner, I made a vertical incision along the length of the spine, slicing through skin and tissue until, going slightly too far, I arrived at the vertebral column itself, announced by the soft knock of the scalpel's steel against bone.

In my beginner's hands, the tapping of the scalpel against the spine sounded like beats of Morse code. The rhythm reminded me of the dean's address to students at the beginning of the term. An infectious-disease specialist, he used a linguistic analogy for medical education, likening our training to the acquisition of a new language. In this way, medical students could be seen as novices sounding their way through volumes of scientific knowledge in search of the vocabulary and grammar of medicine. We would be learning a new alphabet so as to string together words to a story we did not yet know. But not only this. Because language also writes its reader, we would be broken down and reconstituted: rewritten so that we might read the world with new eyes.

But the task at hand was to cut. Carving paths through Stella's flesh, we aimed for symmetry. Radiating perpendicularly from the initial incision along the spine, we made six horizontal cuts. Two bisected Stella's back along the transverse plane. The other four extended outward to points equidistant from her spine. Next, we pulled back thick layers of skin and flesh with tweezers and scissors—teasing them apart from the fascial junction, beneath which glistened rows of muscle fibers. We guided our knives with instructions read aloud from the manual. We peppered one another with questions to reinforce the names of dissected muscles, their nerve supplies, their skeletal origins and insertions. I slowly learned the characteristic feel of each tissue as it was translated through the blade: the leathery resistance of skin, the oily give of fat, the tenacious spring of muscle, the harsh grate of bone. When we finished, broad sections of Stella's back were flayed outward like a set of wings.

Weeks later, we were finishing dissection of the head and neck, the final chapter of our mortal lessons. This last phase required bisection of Stella's skull along the sagittal plane—straight between the eyes—using a vertical band saw that stood as tall as a man. We released the brakes on our anatomy table and wheeled it toward the saw, which sat, like a sturdy red obelisk, in the hall of jarred fetuses, its blade dull and gray and slightly rusted along the edge.

Not much was left of Stella now. Her abdomen had been eviscerated—emptied of bowels, kidneys, stomach, spleen, gallbladder, pancreas, and liver. Her chest wall had been cracked open to get at her heart and lungs. The muscles of her arms and legs were flayed open, just as her back once was. The top of her skull, which we had removed with a hammer and chisel when the bone saw simply couldn't cut it, wobbled like a bowl on the lab bench beside her. Her brain rested in a two-gallon plastic bucket on the floor. We could now study her eyes from the inside of her hollow skull. Parts of her that had begun to grow mold—her foot, a few of her ribs—were lopped off to save what we could.

We transferred what was left of Stella onto the cutting block. Her arms and legs dangled off the sides. With a flip of the switch, one of the professors brought the saw to life, filling the room with a whine. The table shook. The professor gave us the cue, and we slid Stella headfirst toward the blade. Saw met skull. The whine became a high wail and the room filled with the smell of burning bone; a fine ash was thrown up by the cutting—human sawdust.

It was supposed to be easier after the skull. We moved the body farther along the table, letting the saw slice through the forehead, between the eyes, along the nose. It slowly split Stella in two. I focused my gaze just beyond her head, on the white cinderblock wall, tinged slightly yellow and red in places from years of this sacrifice. The final cut through the jaw seemed interminable. The gray blade cut and cut. The professor's hands thrust the jaw closed, breaking a couple of Stella's front teeth.

Then there was a grating metallic screech, the sudden sound of metal biting metal. Sparks erupted from Stella's face. The air filled with the scent of singed flesh as embers spilled from her mouth. "She's on fire!" one of my partners hollered wryly. It certainly seemed that way as the sparks flew. We stepped back from the table—not far, but far enough—leaving the professor alone with Stella as he thrust the saw through her mouth.

"Dentures," he said calmly. The saw did its work. The gray blade cut. Stella spit fire. Finally, like a heavy deadbolt locking, the saw just stopped. I flinched, then beheld the saw stuck motionless, lodged in Stella's jaw. Now it was the jammed machine that shook, trying to break free from Stella's bite.

The professor's collar, always crisply starched, sagged a little around his neck now. He shut off the power and gripped Stella's split face with his hands. Back and forth, he torqued her skull until it was freed from the blade. At last, a plate of artificial teeth landed on the table with a knock, freeing the saw to finish its work.

It was lightning that jolted Victor Frankenstein's creation out of oblivion: a patchwork of cadaveric fragments shook, like Stella trembling in her steel cradle—much like me.

There's no mention of Frankenstein and his monster in the stained pages of my dissection manual, of course—nothing at all that could have prepared me for the spark waiting in that final cut. Perhaps because it wasn't supposed to be so difficult. There should have been no cinders. No struggle. No final metallic screech breaking the spell of ritual. Or perhaps there were simply no words to explain what might happen—because anything could.

A man could enter the cold halls of a cadaver lab and find himself galvanized by the intricacy of the human form, which possesses, even in decay, the beauty of a crumbling temple. There he could become a reluctant disciple, stepping through the temple gates. Carving out a path to its innermost chambers, he could arrive at the altar and ask his burning question: "What have I to do with thee?" And in the sound of his own voice hear himself figured anew.

And so it was. Dismantling the stuff of matter, I had reassembled my mind. Through Stella's unmaking I had made my way closer to becoming a doctor. Where the living raised the dead in Frankenstein's world, the dead animated the living in mine.

In both cases, it's a spark that brought the monster to life.

A few years ago, I asked my old professor if there were any records of Stella's identity in life. "We do not know the names of the cadavers," he replied. "We only know them by the number assigned to them by the State Anatomical Program." When I asked if there was anything else he could tell me, he wrote back: "Female. 90yo. Homemaker. COD Multi-infarct Dementia. That's all the information I have."* When I asked VSAP for additional information, I was told that privacy rules protect the information of decedents for 50 years after the date of death. There is a kind of mercy in this rule; the dead, after all, need their rest.

In the end, my efforts at remembering this woman amount to a kind of eschatological diversion. After all, what is the act of dissection if not a task concerning, quite literally, final matters? If dissection is the study of last things, the paradox of a medical education is that it all begins in the anatomy lab.

When the blade stopped, we slid Stella once more onto our dissection table. I saw the rows of malformed fetuses again: MONSTERS. My reflection stared back at me from the windowpane. Beyond it was a specimen identified only as GIANT: a thickset form bobbing in preservative, its whale-white skin stretched across muscles and bones that hadn't known when to stop growing. Everywhere, osseous formations protruded from under the skin, giving shape to a gnarled, knotted, treelike creature. Its head was huge, an uneven globe perched atop a bulky stalk. The skull looked engorged. The face coiled in upon itself.

Suspended in my memory, this creature has become a reminder of nature's awful power—its cool indifference, its limitless possibilities. I think of those jarred specimens even now, silent witnesses to my education. Each one living up to the etymological meaning of the category: "monster" from the root "monstrum," which gave rise to "demonstrare" and our modern "demonstrate." Indeed, these creatures demonstrated the fragility of life itself: they were signposts at nature's frontier, products of her visionary experimentation. Reminders of what can happen. Of what did.

Our cadavers' remains were consecrated near the university cemetery. Their cremated ashes were scattered in early spring. Around us, the sharp branches of trees rustled in the breeze; tufts of newborn grass marked the green edges of graves. We'd never been like this with them before—outside, in the sunlight. The

* Distinguishing details have been changed to protect Stella's true identity. I am grateful for her immeasurable contribution.

ashes caught the rays as we poured them from plain cardboard boxes. Even here, there were traces of the anatomy lab: in the anonymous numbers identifying the remains, in the rubber gloves worn by students dispersing ashes, in the care we took to keep from mingling with the dust. After a box full of ash was deposited in the cemetery, the remaining boxes were driven to the summit of the university's astronomical observatory. And though I shall never quite know her, it is here, scattered on the mountaintop, that I like to think Stella found her final rest. And it is here, when I want to seek her out again, that I'll come looking for the stardust under my boots.

KATIE WORTH

Telescope Wars

FROM *Scientific American*

FOR 15 YEARS three competing groups of astronomers have chased a single dream: to build the grandest telescope on earth. The stargazing behemoths they aim to build would be three times larger than the world's current largest optical telescopes, powerful enough to take pictures of planets circling other stars and to peer across the breadth of the universe, gazing back in time nearly to the Big Bang.

This dream observatory comes in three versions: the Giant Magellan Telescope (GMT), developed by a consortium including the Carnegie Institution for Science; the Thirty Meter Telescope (TMT), developed by the California Institute of Technology, the University of California system, and others; and the European Extremely Large Telescope (E-ELT), developed by the European Southern Observatory (ESO). Building all three would cost nearly $4 billion, but so far the world has balked, leaving each project short on cash and hustling for more. There could have been at least one giant telescope gazing at the heavens today; instead, partially built hardware awaits delivery to barren construction sites.

All three telescopes are likely to limp across the finish line of their race and begin operations sometime in the 2020s, albeit behind schedule and over budget.

How did this happen? How did three separate projects with common goals come to be fighting one another for funding? And what has prevented them from joining forces to minimize the chance of their collective failure?

These questions have been asked repeatedly, including by a

bewildered national panel considering two of the telescopes for federal funding. Dozens of scientists interviewed for this story pondered what might have been if instead of three ventures, there had been one or two. Nearly all agreed that humankind would be much closer to building the next, greatest generation of observatories if competing groups of astronomers had not spurned repeated chances to collaborate. That competition started in the first decades of the 20th century and has been sustained across the years by personality conflicts, miscommunications, competing technologies, and an expanding universe of bitterness.

The Big Deal

The story begins in 1917, when an ambitious astronomer and observatory director named George Ellery Hale unveiled something entirely new to science, a 100-inch optical telescope.

In the world of telescope construction, size matters: the larger a telescope's mirror, the farther it sees. The new telescope, perched on Mount Wilson in what was then still a dark-skied Los Angeles County, dwarfed all others on earth. Its revolutionary size rapidly produced revolutionary results. Edwin Hubble used it to discover that our galaxy is but one among many and then to gather evidence that the universe is expanding.

But Hale was not satisfied. He wanted a 200-inch telescope.

The 100-inch one was built and run by what was then called the Carnegie Institution of Washington, a charity created by steel baron Andrew Carnegie. Carnegie was not prepared to spend millions more on a new telescope, so Hale slyly pitched the project to an organization funded by Carnegie's rival, oil magnate John D. Rockefeller. In 1928 Rockefeller personally approved Hale's 200-inch telescope, eventually providing it with a $6 million grant—at the time, the largest sum ever donated to a scientific project.

There was a catch: the astronomers at the Carnegie Institution were the only ones in the world with the expertise to build the new telescope, but Rockefeller would not fund his old rival's charity. "It was just not going to happen," says historian Ronald Florence, who wrote *The Perfect Machine*, a book about the 200-inch telescope. "So that sets up the pool shot for problems."

Hale came up with a solution: Rockefeller would give the telescope money as a gift to Caltech, which had just been established only two miles (three kilometers) from Carnegie's observatory headquarters in Pasadena, California. Caltech was still so embryonic that it did not employ a single astronomer, let alone an astrophysics department. Nevertheless, the Rockefeller Foundation funded Caltech's construction of Hale's new telescope and the Palomar Observatory in San Diego County, which housed it. Hale believed Carnegie's leaders would find working on such a magnificent stargazing tool irresistible and would lend their expertise to design and construct the new telescope.

Hale was mistaken. According to Florence, the deal enraged the Carnegie Institution's president, John Merriam, who saw it as an unforgivable deceit. He worked to scuttle the project, refusing to allow Carnegie scientists to help and pressuring the Rockefeller Foundation to walk away. Desperate, Hale called on the diplomat Elihu Root, an old friend of both Rockefeller and Carnegie's. Root swayed Merriam, who at last signed on to the project.

But the discord was only beginning: Merriam was still angry and tried for years to wrest control from Caltech, Florence says, until the institutional distrust became mutual and profound.

After Merriam retired, the warring charities at last formed an uneasy truce. The Rockefeller Foundation approached its astronomical adversaries with a deal: Caltech would own the telescope when it opened its 16-foot eye in 1949, but Carnegie would operate it.

The fragile relationship between the institutions inevitably spilled into science, especially after the identification of "quasi-stellar objects"—quasars—in the early 1960s by Dutch American astronomer Maarten Schmidt. Although they at first seemed to be dim stars in the sky, further studies showed quasars to be shining with almost unthinkable brilliance from the far-distant universe. The mysterious objects quickly became astronomy's sexiest subject, and Caltech and Carnegie researchers vied for time on the world's largest telescope to study them, sometimes resorting to "junior high–level pettiness," Florence says.

In 1979, after half a century of tensions, Caltech finally sought to end its strained shared custody of Palomar. The split did not go well and proved intensely personal. The late Allan Sandage,

Carnegie's legendary astronomer, who had achieved his life's work at Palomar, refused to set foot in the observatory again. "It was the kind of divorce where you had to choose the husband or the wife," Florence says. "There was no staying friends with both."

Conflicting Designs

Over the next two decades the institutions trod separate paths. In the 1990s Caltech partnered with the University of California to unveil the twin 10-meter Keck telescopes on Mauna Kea in Hawaii, using what was then a novel segmented-mirror design in which many small mirrors created one larger, light-gathering aperture. Their risk paid off: the design worked beautifully, and their astronomers enjoyed years of scientific preeminence before anyone else built something competitive. Meanwhile Carnegie stuck with the older, single-mirror technology but ventured into the Southern Hemisphere, constructing the twin 6.5-meter Magellan telescopes in the Atacama Desert in northern Chile.

Carnegie was just completing these telescopes in 1999 when Caltech and the University of California announced their intention to build a 30-meter telescope. The ESO, an intergovernmental organization of astronomers throughout Europe, was already toying with something even more ambitious—a 100-meter (and appropriately named) OverWhelmingly Large Telescope.

To most astronomers, jumping from a 10-meter telescope to a 100-meter one was absurdly ambitious. But a 30-meter telescope seemed viable, to the consternation of Gus Oemler, then the observatories director at Carnegie. He remembers waking up to Caltech's announcement and feeling sick. "We were struggling to finish the Magellan telescopes, which were finally going to give us some kind of parity with Caltech after many years, and suddenly they were starting the next phase."

After much debate, Carnegie pitched Caltech on a collaboration. Both sides were hesitant, but the boards of each institution thought it was time to traverse the freeway and the old grudge that separated them. "We recognized it would be kind of crazy to have two giant telescopes centered on two institutions within two miles of each other," says Carnegie astronomer Alan Dressler.

So on June 21, 2000, two scientists from Caltech—the late astronomer Wal Sargent and the late Tom Tombrello, then the physics chair—and two from Carnegie—Oemler and Dressler—met to discuss a partnership.

By all accounts, that discussion went terribly. The meeting was tense, disjointed, and plagued by misunderstanding. Both Wendy Freedman, who would later become director of the Carnegie Observatories, and Richard Ellis, now a senior scientist at the ESO, who was then on the verge of replacing Sargent as Caltech's Palomar Observatory director, spoke to all four men immediately after the meeting and heard a different story from each: Dressler felt that the Caltech men were not taking Carnegie's proposal seriously, whereas Tombrello mistakenly believed that Carnegie did not have serious money to contribute. Oemler said Sargent sat in icy silence through most of the meeting. Sargent later said he was worried about upsetting Caltech's then-delicate relationship with the University of California. But Sargent had not explained that concern during the meeting, Ellis says, so "of course, the Carnegie people were offended."

The next day Tombrello sent an email "to summarize our rambling discussion." Caltech was not interested in working with Carnegie on the telescope for the time being, Tombrello wrote, although he would not exclude the possibility if the work got expensive. The Carnegie astronomers felt condescended to and insulted. The nascent collaboration died, and the long tradition of acrimony between the institutions grew longer.

That meeting is now a part of giant-telescope lore. Ellis is one of many astronomers who wonder what might have happened had the meeting gone differently.

"When you look back on that moment—what a tragedy," he says. "With a few phone calls and a bit of diplomacy, we could have brought Carnegie in. And had we brought them in, we'd probably have a telescope by now."

Garth Illingworth, an astronomer at the University of California, Santa Cruz, says there remained "just enough residual resentment and unhappiness" from the old rivalry to derail a constructive conversation. "You just think, jeez, why wasn't there a little adult supervision in the room to help these folks to get over this?" he adds.

Divided They Fall

After this failed détente, the rivalry only expanded. Caltech and the UC system developed the TMT, to be constructed next to the Keck telescopes in Hawaii. Meanwhile Carnegie designed the GMT, a 24.5-meter telescope, to cap its Las Campanas Observatory in Chile. Around the same time, the Europeans scaled down their dreams from overwhelmingly large to merely extremely large and planned the construction of the 39-meter E-ELT in Chile.

The three projects scoured the globe for financing, sometimes searching in the same places. Pony up money, the typical pitch went, and your astronomers will be guaranteed telescope time. Canadian astronomers, for instance, were courted by both the Carnegie group and the Caltech-UC teams and chose the latter. Harvard University was also wooed by both but committed to Carnegie. At least once, the two American teams awkwardly ran into each other in an airport as they traveled to meetings with the same potential partners. And the Europeans were not above the fray: they initially secured support from Brazil, whose president agreed to join the ESO and underwrite a major chunk of the E-ELT. But fractured Brazilian politics stalled the agreement. Carnegie has taken advantage of the E-ELT's woes: in July 2014 the University of São Paulo joined the GMT project, and according to Dressler, GMT leadership hoped the Brazilian government would soon follow, although that has not turned out to be the case.

The most sought-after partner of all has been the U.S. government, which could open its strongbox of federal funding to finance a giant telescope and provide access for all American astronomers. In 2000 the Astronomy and Astrophysics Decadal Survey, a once-a-decade national panel that guides U.S. federal funding, had declared a next-generation giant telescope the country's highest priority in ground-based optical astronomy.

With this endorsement, the National Science Foundation began discussing a partnership with the Caltech-UC TMT project in 2003. But within months GMT astronomers wrote a letter saying the deal would unfairly favor the TMT. The letter was effective: the NSF balked, unwilling to take sides in the increasingly divisive politics of top-tier optical astronomy.

In reality, there was not much federal money to provide anyway, according to NSF senior adviser Wayne Van Citters. But the feud did not help, he says: "We needed the community to come together and decide which one they wanted to do. We couldn't possibly do both."

The community, for its part, tried repeatedly to do just that, but the efforts proved fruitless. European astronomers discussed collaborations with both their rivals but ultimately only agreed to share technology insights. And in 2007, at the insistence of their boards, TMT and GMT leaders held several coldly cordial meetings to discuss ways they might work together. Nothing came of it.

The situation confounded panel members of the 2010 decadal survey, who questioned why the U.S. astronomy community was being asked to support two separate American-led large optical telescopes. In the end, they backed neither, kicking the projects to the bottom of the priority list and effectively quashing federal funding for another 10 years.

Rivalry is hardly rare in science: brilliant minds are often accompanied by big egos with a penchant for clashing. Sometimes feuds can yield innovation; other times they can turn the high-minded pursuit of discovery into a series of petty personal conflicts. Some disciplines have successfully convinced potential rivals to join forces: High-energy physicists work in massive international ensembles on particle accelerators. Radio astronomers have collaborated on their field's largest next-generation tool, the $1.4-billion Atacama Large Millimeter/submillimeter Array.

In contrast, optical astronomy in the United States has been riven with competition. Italian American astronomer and Nobel laureate Riccardo Giacconi described it in a July 2001 speech to the National Academy of Sciences as a sociological problem.

To historian W. Patrick McCray of the University of California, Santa Barbara, who wrote *Giant Telescopes,* a book about the American optical-astronomy community, what is striking about the enmity between Caltech and Carnegie is its longevity: they have been bickering over large telescopes since 1928. "You just think, Have you people learned nothing?" McCray says.

But rivalry alone does not explain the state of affairs. There were rational reasons to work on separate telescopes, notes astronomer Ray Carlberg of the University of Toronto, which is part of an association involved with the TMT project. Initially astronomers believed there would be money for all three, and giant telescopes in both the Northern and Southern Hemispheres would ensure full coverage of the entire sky. "The world had just built quite a few eight- and ten-meter telescopes, and it didn't seem unreasonable to have a bunch of these big ones," Carlberg says. By the time it was clear that Caltech could use Carnegie's help, Carnegie was too deeply invested in its own project to abandon it.

Too Many Telescopes

On the Big Island of Hawaii, a corner of Mauna Kea's immense summit has been flattened to make way for the TMT. The telescope's 30-meter mirror, the diameter of the U.S. Capitol dome, will be composed of a honeycomb of 492 hexagonal, 1.44-meter segments, all housed in an 18-story structure on the dormant volcano. The project has been granted land-use permits, although it still faces vocal opposition and legal challenges from some native Hawaiians and environmentalists. To help pay for the $1.5-billion endeavor, Caltech and the UC system have secured international partnerships with India, China, Japan, and Canada. They are still searching for an additional $270 million; the project's current best guess for its telescope's debut is sometime in the early 2020s.

Eleven blocks from the TMT's Pasadena headquarters, Carnegie and its partners are coaxing the 24.5-meter GMT into life. It will consist of seven 8.4-meter mirrors, with six mirrors arranged like flower petals around one in the center—an approach very different from, and incompatible with, the smaller, more numerous hexagonal mirrors of the TMT. Four mirrors have already been cast at a laboratory at the University of Arizona. The more modest size and design come with a more modest cost: just under $1 billion. Carnegie has enlisted the support of universities from South Korea, Australia, and Brazil, as well as several domestic universities. They have raised roughly half the money needed to build the telescope at its construction site within the Las Campanas Observatory. If all goes as planned, the GMT will begin collecting light by 2022.

A 12-hour drive up the Pan-American Highway from Las Campanas is Cerro Armazones, the desert mountain where the E-ELT will one day perch. The site was initially scoped out by TMT astronomers, who spent years monitoring the atmosphere above Cerro Armazones for transparency and turbulence before concluding they preferred to build in the Northern Hemisphere instead; the Europeans took advantage of that groundwork and claimed Armazones for their own project. Today a newly paved road leads to the mountain's bald scalp, which has been shaved with dynamite and heavy machinery into a soccer-field-sized flattop. Visible to the east of the mountain, the firmament meets the 6,723-meter Andean volcano Llullaillaco, where the Inca once sacrificed children to the gods. It and the rest of the arid panorama fade at nightfall, making way for a playground of stars overhead.

With a mirror 39 meters wide, the E-ELT will be the grandest next-generation telescope of all. Like the TMT, the E-ELT will have a segmented design, but instead of 492 hexagonal mirrors, it will boast 798. In December 2014 the ESO voted to move forward with first-phase construction. A second phase has not yet been funded. The E-ELT's leadership plans for the telescope to begin stargazing in 2024, for a total construction cost of €1.1 billion.

Once constructed, the three telescopes will have synergistic strengths, says the E-ELT's Roberto Gilmozzi. The E-ELT will specialize in zooming in to provide high-resolution images of small regions of the sky; the GMT will excel at wide-field astronomy. And the TMT will be located in a different hemisphere, observing a different sky.

Gilmozzi, like most other astronomers interviewed for this story, thinks that had there been two telescopes instead of three, both might be nearing completion by now, at a cost hundreds of millions of dollars less. "If you don't consider the problem of finding the money, it's wonderful to have more than one," he says. "Scientifically speaking, I could use 100 telescopes if I could afford to build them."

Unfortunately, building telescopes is just the first step. Neither the GMT nor the TMT currently has enough money to sustain operations once it is constructed. Both hope the federal government will eventually step in to assist, but Van Citters says it is not clear how much money the government will be able to contribute. The telescopes are each expected to cost tens of millions of dollars a year to operate. "That's enough to give people nightmares," Mc-Cray says.

Even so, the problem of too many telescopes has a silver lining: the world could one day have three giant eyes gazing at the cosmos. This would be a big win for science, McCray says. "If this situation is a tragedy, it's a tragedy with a small *t*."

Contributors' Notes

Other Notable Science and Nature Writing of 2015

Contributors' Notes

Chelsea Biondolillo is the author of the prose chapbooks *Ologies* (2015) and *#Lovesong* (2016). Her essays and journalism have been published widely online and in print. She teaches, writes, and hikes in Phoenix, Arizona.

Bryan Christy, a lawyer and former CPA, is the founder of the Special Investigations Unit at *National Geographic.* He is the author of *The Lizard King: The True Crimes and Passions of the World's Greatest Reptile Smugglers.* In 2014 he was named the National Geographic Society's Explorer of the Year.

Helene Cooper is the Pentagon correspondent for the *New York Times,* where she shared a 2015 Pulitzer Prize for International Reporting for her work covering the Ebola crisis in West Africa. From 2008 to 2012 she covered the White House, and prior to that she was the *Times*'s diplomatic correspondent. She joined the newspaper in 2004 as the assistant editorial page editor, a position she held for two years before she ran out of opinions and returned to news. She has reported from 64 countries, from Pakistan to the Congo. Born in Monrovia, Liberia, Cooper is the author of *The House at Sugar Beach: In Search of a Lost African Childhood,* a *New York Times* bestseller and a National Book Critics Circle finalist in autobiography in 2009.

Gretel Ehrlich is the author of 15 books of nonfiction, fiction, and poetry, including *The Solace of Open Spaces, Heart Mountain, This Cold Heaven,* and *Facing the Wave,* longlisted for the National Book Award. Her books have won many awards, including the first Henry David Thoreau Prize for nature writing and the PEN Center USA Award for Creative Nonfiction. She is the recipient of a Guggenheim Fellowship, three National Geographic Expeditions Council grants for travel in the Arctic, a Whiting Award, and

National Endowment for the Arts and National Endowment for the Humanities grants. Her poetry was featured on *PBS NewsHour.* She lives with her partner, Neal Conan, on a farm in the highlands of Hawaii and a ranch in Wyoming.

Rose Eveleth is a journalist who covers how humans tangle with science and technology. She's the host and producer of *Flash Forward,* a podcast about the future, and has covered everything from fake tumbleweed farms to sexist prosthetics.

Amanda Gefter is an award-winning science writer specializing in fundamental physics and the philosopy of science, and the author of *Trespassing on Einstein's Lawn* (2014). Her articles have been featured in *Nautilus, Scientific American, Quanta, Nature, NOVA Next, BBC Earth, The Atlantic, Technology Review,* and *New Scientist.* She was a 2012–2013 Knight Science Journalism Fellow at MIT and lives in Somerville, Massachusetts.

Rose George is an author and journalist based in Yorkshire, England. She has written three books, on refugees, sanitation, and shipping, and her articles have appeared in the *Guardian, New Statesman, Scientific American,* and many other publications. She is working on a book about blood. When she's not doing that, she can usually be found running up and down hills.

Gabrielle Glaser is an award-winning journalist whose work on mental health and medical and social trends has appeared in *The Atlantic,* the *New York Times Magazine,* the *New York Times,* the *Daily Beast,* the *Washington Post,* the *Los Angeles Times, Health,* and many other publications. She is the author of three books, most recently *Her Best-Kept Secret: Why Women Drink—and How They Can Regain Control,* a *New York Times* bestseller. She received bachelor's and master's degrees in history from Stanford University.

Antonia Juhasz is an energy analyst, author, and freelance investigative journalist specializing in oil. Her writing appears regularly in numerous publications, including *Harper's Magazine, Newsweek, Rolling Stone, The Atlantic, The Nation, Ms., The Advocate,* CNN.com, the *New York Times,* and the *Los Angeles Times.* She is a contributing author to six books and the author of three: *Black Tide: The Devastating Impact of the Gulf Oil Spill* (2011), *The Tyranny of Oil* (2008), and *The Bush Agenda* (2006). Her investigations have taken her a mile below the surface of the Gulf of Mexico and to the rainforests of the Ecuadorean Amazon, from the deserts of Afghanistan to the fracking fields of North Dakota, and to the tip of the Alaskan Arctic. She holds bachelor's and master's degrees in public policy from Georgetown and Brown Universities, respectively. She was a fellow of the Investigative Reporting Program at the University of California, Berkeley.

Alexandra Kleeman was raised in Colorado and has been published in *The Paris Review, Zoetrope: All-Story, Harper's Magazine, n+1, The New Yorker,* and *the Guardian.* Her work has received scholarships and grants from Bread Loaf, the Virginia Center for the Creative Arts, and the Santa Fe Art Institute. She is the author of the novel *You Too Can Have a Body Like Mine* (2015) and *Intimations* (2016), a short story collection. She lives in Staten Island and is the 2016 recipient of the Bard Fiction Prize.

Elizabeth Kolbert is a staff writer for *The New Yorker* and the author of *The Sixth Extinction,* which won the 2015 Pulitzer Prize. Her three-part series on global warming, "The Climate of Man," won the American Association for the Advancement of Science's Journalism Award and a National Academies Communication Award. Those articles became the basis for *Field Notes from a Catastrophe: Man, Nature, and Climate Change.* She is a two-time National Magazine Award winner and has received a Heinz Award and a Lannan Literary Fellowship. Kolbert lives in Williamstown, Massachusetts.

Kea Krause has written for *The Believer, The Toast, Vice, Narratively,* and *The Rumpus.* She holds an MFA from Columbia University, where she also taught creative writing. Born and raised in the Pacific Northwest, she now resides in Queens, New York.

Robert Kunzig is the senior environment editor for *National Geographic.* His article on cities was the last in the magazine's yearlong series on global population trends. A science journalist for more than 30 years, including 14 at *Discover,* Kunzig has written two books, *Fixing Climate* (with Wallace Broecker) and *Mapping the Deep,* about oceanography, which won the Aventis Prize for General Science Book, the Royal Society science book prize, in 2001. He lives in Birmingham, Alabama, with his wife, Karen Fitzpatrick.

Amy Leach's work has been published in *A Public Space, Tin House, Orion, The Los Angeles Review,* and many other places. She has been recognized with a Whiting Award, selections in *The Best American Essays,* a Rona Jaffe Foundation Writers' Award, and a Pushcart Prize. She plays bluegrass, teaches English, and lives in Montana. *Things That Are* is her first book.

Apoorva Mandavilli is founding editor and editor in chief of *Spectrum.* She is also an adjunct professor in New York University's Science, Health and

Environmental Reporting Program, and cocreator of Culture Dish, an organization that aims to enhance diversity in science writing. Her work has appeared in *The New Yorker, The Atlantic, Slate, Nature,* and *Popular Science,* among others.

Charles C. Mann's most recent book, *1491,* won the U.S. National Academy of Sciences' Keck Award for best book of the year. Mann is at work on a new project, a book about the future that makes no predictions.

Emma Marris is a writer based in Klamath Falls, Oregon, not far from the territory of the wolf called OR7. She writes about people, nature, food, culture, and science.

Sarah Maslin Nir has been a staff reporter for the *New York Times* since August 2011. She currently covers Brooklyn for the paper's metro section. Prior to that, she covered Manhattan and Queens and worked as a rewrite reporter for late-night breaking news.

Maddie Oatman writes and edits stories about food, culture, and the environment for *Mother Jones.* She also cohosts the magazine's food and agriculture podcast, *Bite.* Her work has appeared in *Grist,* the *Huffington Post, Outside,* and *The Rumpus,* and she was a 2013 Middlebury Fellow in Environmental Journalism. A Colorado native now based in San Francisco, she makes time for backcountry skiing and exploring the outdoors with friends.

Stephen Ornes writes about mathematics, physics, space, and cancer research from his home in Nashville, Tennessee, where he lives with his wife and three children. "The Whole Universe Catalog," which appears here, was a story three years in the making. His first book, published in 2008, was a young-adult biography of mathematician Sophie Germain, and he contributed to *The Science Writers' Handbook,* published in 2013. In addition to *Scientific American,* his work has appeared in *New Scientist, Discover, Science News for Students, Physics World,* and *Cancer Today.* He teaches a class in science communication at Vanderbilt University.

Rinku Patel is writing a book about immune dysfunction and holistic medicine.

Oliver Sacks, who died in 2015, was a physician and the author of many books, including *The Man Who Mistook His Wife for a Hat, Musicophilia,* and *Hallucinations.* His book *Awakenings* has inspired a number of dramatic adaptations, including a play by Harold Pinter, an Oscar-nominated film, and a documentary by Bill Morrison with music by Philip Glass. Dr. Sacks was a professor of neurology at the New York University School of Medicine and a visiting professor at Warwick University in the United Kingdom.

Kathryn Schulz is a staff writer for *The New Yorker.*

Gaurav Raj Telhan is a physician and writer. He studied literature and medicine at the University of Virginia, where he was a Crispell Scholar. He was chief resident at NYPH/Columbia-Cornell University Medical Center and trained at Memorial Sloan Kettering Cancer Center. He is a recipient of the Smith-Shanubi Scholarship at the New York State Writers Institute. Dr. Telhan practices interventional spine medicine in an academic medical center. He writes about the intersection of literature, philosophy, and medicine.

Katie Worth is a journalist who writes about science and politics and their myriad intersections. She has covered the reason why the seemingly settled science of DNA forensics may be anything but, the role of the WhatsApp messaging platform in Brazil's response to Zika, and the psychology of the fight over climate change. Her stories have been published in *Scientific American, National Geographic News, Vice,* the *Wall Street Journal,* and *Slate.* She is now a digital reporting fellow at *Frontline,* a PBS investigative-documentary outlet based at WGBH in Boston. Her reporting for "Telescope Wars" took place while she lived in Chile, home to many of the world's largest telescopes.

Other Notable Science and Nature Writing of 2015

SELECTED BY TIM FOLGER

JAKE ABRAHAMSON
To the Sea and Back. *Sierra,* January/February.
JOEL ACHENBACH
The Age of Disbelief. *National Geographic,* March.
CAROLINE ALEXANDER
The Invisible War on the Brain. *National Geographic,* February.
GUSTAVE AXELSON
Soul Mates. *Living Bird,* Autumn.

MICHAEL BALTER
Flight School. *Audubon,* January/February.
BELLE BOGGS
Baby Fever. *Orion,* November/December.
JAMES BOYCE
Rethinking Extinction. *Harper's Magazine,* November.
MIKE BROWN
Why I Still Love Pluto. *Popular Science,* July.

JOHN COLAPINTO
Lighting the Brain. *The New Yorker,* May 18.
KRISTA CONGER
The Butterfly Effect. *Stanford Medicine,* Summer.
GARETH COOK
The Singular Mind of Terry Tao. *The New York Times,* July 24.

TIMOTHY FERRISS
First Glimpse. *National Geographic,* January.

CHARLES FISHMAN
5,200 Days in Space. *The Atlantic,* January/February.
DOUGLAS FOX
Life at Hell's Gate. *Scientific American,* July.
IAN FRAZIER
Earle Power. *Outside,* November.
DAVID FREEDMAN
Birth of a Rocket. *Scientific American,* June.

BRIAN GREENE
Why He Matters. *Scientific American,* September.

ERIKA CHECK HAYDEN
Bloodline. *Wired,* June.
BERND HEINRICH
Chickadees in Winter. *Natural History,* March.

WALTER ISAACSON
How Einstein Reinvented Reality. *Scientific American,* September.

WES JACKSON
The Great Awakening. *Catamaran Literary Reader,* Spring.
ROWAN JACOBSEN
The Perfect Beast. *Outside,* January.
OLIVIA JUDSON
Luminous Life. *National Geographic,* March.
RUSS JUSKALIAN
Climate and the Khan. *Discover,* July/August.

PAUL KALANITHI
Before I Go. *Stanford Medicine,* Spring.
SAM KEAN
The Man Who Couldn't Stop Giving. *The Atlantic,* May.
PAUL KVINTA
Overroo'd!. *Outside,* December.

WINONA LADUKE
The Thunderbirds Versus the Black Snake. *Earth Island Journal,* Autumn.
MICHAEL LEMONICK
The Archaeology of Stars. *Discover,* December.

LINDA MARSA
The People's Scientist. *Discover,* November.
MAC MCCLELLAND
Slip Sliding Away. *Audubon,* March/April.
TRISTAN MCCONNELL
The End for Elephants?. *Earth Island Journal,* Summer.

Bill McKibben
Power to the People. *The New Yorker,* June 29.
Kent Meyers
The Quietest Place in the Universe. *Harper's Magazine,* May.
Steve Mirsky
Great Feets. *Scientific American,* November.
Virginia Morell
Feeding Frenzy. *National Geographic,* July.
George Musser
Is the Cosmos Random?. *Scientific American,* September.

Nick Neely
Discovering Anna. *The Southern Review,* Winter.
Michelle Nijhuis
Harnessing the Mekong. *National Geographic,* May.

Michael Pollan
The Trip Treatment. *The New Yorker,* February 9.
Corey Powell
Europa or Bust. *Popular Science,* September.

John Richardson
Ballad of the Sad Climatologists. *Esquire,* August.
Michelle Robertson
Recovery Season. *Orion,* September/October.

Oliver Sacks
A General Feeling of Disorder. *The New York Review of Books,* April 2.
Carl Safina
Big Love. *Orion,* May/June.
Michael Schulson
In Evangelical Country, an Apocalypse of Rising Seas. *Insideclimatenews.org,* December 8.
Jamie Shreeve
Mystery Man. *National Geographic,* September.
Hampton Sides
High Science. *National Geographic,* June.
Tom Siegfried
The Amateur Who Helped Einstein See the Light. *Science News,* October 1.
Lauren Slater
Bloodlines. *The Sun,* March.
Jason Smith
Guest of the Great White. *Catamaran Literary Reader,* Summer.
Gary Stix
Lifting the Curse of Alzheimer's. *Scientific American,* May.
Abe Streep
Coal. Guns. Freedom. *The California Sunday Magazine,* October 13.

NANDHINI SUNDARESAN
Catch and Release. *Harvard Review* 46.

LEATH TONINO
In Pursuit of Bird Poop. *Tricycle.com*, July 28.
BIJAL TRIVEDI
Broken Cables. *Discover*, September.

ERICK VANCE
An Ocean Apart. *VQR*, Spring.

W. D. WETHERELL
Outage. *Appalachia*, Summer/Fall.
DEBRA WHITE PLUME
Of Water, Starlight, and Raspberries. *Earth Island Journal*, Autumn.
ELLIOTT WOODS
Wolflandia. *Outside*, February.

TIM ZIMMERMANN
The Piscivore's Dilemma. *Outside*, June.